Atomic Force
Microscopy

Atomic Force Microscopy
Fundamental Concepts and Laboratory Investigations

Wesley C. Sanders

CRC Press
Taylor & Francis Group
Boca Raton London New York

CRC Press is an imprint of the
Taylor & Francis Group, an **informa** business

CRC Press
Taylor & Francis Group
6000 Broken Sound Parkway NW, Suite 300
Boca Raton, FL 33487–2742

© 2020 by Taylor & Francis Group, LLC

CRC Press is an imprint of Taylor & Francis Group, an Informa business

No claim to original U.S. Government works

Printed on acid-free paper

International Standard Book Number-13: 978-0-367-21864-5 (Paperback)
978-0-367-37123-4 (Hardback)

Visit the Taylor & Francis Web site at
www.taylorandfrancis.com

and the CRC Press Web site at
www.crcpress.com

eResource material is available for this title at https://www.crcpress.com/9780367371234.

Contents

Preface

Nanotechnology is a scientific field involving the study and synthesis of materials with one dimension 1 to 100 nanometers in size. The unique optical, electrical, chemical, and mechanical properties of nanomaterials are apparent when the physical dimensions fall in this range. Optical microscopy techniques are insufficient for the characterization of nanomaterials. The resolution limit for optical microscopes is approximately 200 nanometers. For this reason, more advanced characterization techniques are imperative. Scanning probe microscopy is one of the most common tools in surface science. Scanning probe microscopes possess the ability to measure properties of material, chemical, and biological surfaces. Scanning probe microscopes, particularly the atomic force microscope, are now ubiquitous in research laboratories throughout the world. The atomic force microscope is widely regarded as the technique responsible for ushering in the study of nanoscale materials. This book focuses primarily on the atomic force microscope, and serves as a reference for students, postdocs, and researchers using atomic force microscopes for the first time. In addition, this book can serve as the primary text for a semester long, introductory course in atomic force microscopy. There are a few algebra-based mathematical relationships included in this book that describe the mechanical properties, behaviors, and intermolecular forces associated with probes used in atomic force microscopy. Chapter 1 provides a general overview of scanning probe microscopy. Chapter 2 describes the tip-sample forces that are present during various operating modes of atomic force microscopy. In Chapter 3, readers will get an overview of the basic operation of atomic force microscope electronics. Key components described in Chapter 3 include photodiodes, scanners, and the feedback loop. In addition, Chapter 3 describes the importance of key parameters used to acquire high-resolution images, particularly the proportional and integral gains. Chapter 4 provides information regarding characteristic properties associated with probes used in atomic force microscope applications. The remaining chapters address the basic operating modes of the atomic force microscope. Chapter 5 covers contact mode. Chapters 6 and 7 cover frictional force microscopy and conductive atomic force microscopy respectively. Both of these operating modes are subsets of contact mode atomic force microscopy. Chapter 8 is an overview of oscillating modes of atomic force microscopy, such as tapping mode, non-contact mode, and nanoscale impedance microscopy. A list of key objectives is at the start of each chapter. In addition, relevant figures, tables, and illustrations are in each chapter in an effort to provide additional information and interest. This book includes suggested laboratory investigations that provide an opportunity to explore the versatility of the atomic force microscope. These laboratory exercises include opportunities for experimenters to explore force curves, surface roughness, friction loops, conductivity imaging, and phase imaging. Readers will find questions at the end of each chapter to reinforce laboratory and chapter content. References are included in each chapter to provide opportunities for further reading.

Wesley C. Sanders

Author

Wesley C. Sanders is currently an associate professor at Salt Lake Community College. He teaches courses in nanotechnology, materials science, chemistry, and microscopy. He has published articles in the *Journal of Chemical Education* describing undergraduate labs involving the use of the atomic force microscope. He earned a BSEd in science education from Western Carolina University (1999). Later, he earned a MS in chemistry from the University of North Carolina at Charlotte (2005) and a PhD in chemistry from Virginia Tech (2008). His initial experiences with microscopy involved the study of self-assembled monolayers on gold with a scanning electrochemical microscope as a doctoral student at Virginia Tech. After receiving his PhD, he examined bacterial nanofilaments with an atomic force microscope while working as a postdoctoral researcher at the US Naval Research Laboratory in Washington, DC.

Acknowledgments

I would like to thank former and current microscopy lab facilitators at Salt Lake Community College, Sam Lindsey, Gabe Glass, and Glen Johnson, in addition to several former students who were instrumental in the testing and implementation of the laboratory exercises described in this text.

1

Introduction to Atomic Force Microscopy

1.0 Key Objectives

To obtain a general overview of atomic force microscope:

- operation
- components
- applications
- probe characteristics
- tip-sample forces

1.1 Scanning Probe Microscope Overview

Scanning probe microscopes (SPMs) are a class of microscopes that capture surface topography using probes that scan sample surfaces [1]. SPM is one of the most common tools in surface science [2]. SPMs are now available in several research labs throughout the world and are widely regarded as the technique that ushered in the study of matter at the nanoscale [3]. SPMs do not use glass or magnetic lenses for producing images [4]. Image acquisition involves scanning across sample surfaces with a sharp probe that monitors tip-sample interactions to generate images [5]. The two primary forms of SPM are the scanning tunneling microscope (STM) and the atomic force microscope (AFM) [1]. These techniques allow extreme magnifications in the x-, y-, and z-directions, which facilitate atomic scale imaging with high resolution. These instruments can be used in any environment, such as ambient air, various gases, liquids, vacuum, low temperatures, and high temperatures. Imaging in liquid allows the study of live biological samples and eliminates the capillary forces that are present at the tip-sample interface when imaging samples in ambient air [6]. SPM origins involve the invention of the STM in 1981 at IBM (Zurich) by

Heinrich Rohrer and Gerd Binning. STMs use an electrical tunneling current between a metal tip and sample to record sample topography (Figure 1.1). Although the ability of the STM to image surfaces with atomic resolution caused a great impact on the technology community, STM imaging is limited to conductive materials [1].

The need to study other materials led to the development, in 1986, of the AFM by Gerd Binning, Calvin Quate, and Cristoph Gerber. The AFM enabled detection of atomic scale features on a wide range of insulating surfaces [1]. AFM inventors replaced the wire of a tunneling probe from the STM with a lever made by carefully gluing a tiny diamond onto the end of a spring made of a thin strip of gold. This was the cantilever of the first AFM. This made acquisition of extremely high-resolution images of nearly any sample possible [7]. AFMs are capable of investigating the surfaces of both conductors and insulators on an atomic scale if suitable techniques for measuring cantilever motion are used [6]. AFMs measure ultrasmall forces between the AFM tip and the sample surface. Users can operate AFMs in lateral force microscope (LFM) mode; this mode possesses the ability to measure both normal and lateral forces [6]. AFMs produce images, almost at the level of atomic resolution, by measuring the contour of the sample. Image acquisition involves quantifying the forces

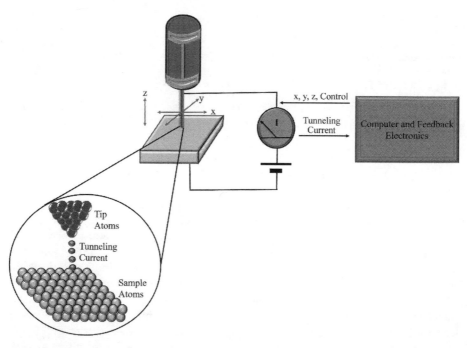

FIGURE 1.1
STM components and operation.

between the AFM tip and the sample surface. Initially, AFM use was limited solely topographical imaging; later AFM use expanded to biological samples [5]. AFMs have become a widespread technology based on the use of micro- and/or nano-structured probes for scanning materials surfaces with sub-nanometer and atomic resolution [8].

1.2 AFM Description

AFM is one of the most versatile techniques for surface analysis. This method allows users to see the shape of a surface in three-dimensional (3D) detail down to the nanometer scale. AFM can image all materials—hard or soft, synthetic or natural—regardless of opaqueness or conductivity [9]. Monitoring one or more interactions with AFM unveils the surface morphology, surface and subsurface organization, and/or physical and chemical properties of the materials under investigation with unprecedented nanoscale lateral and vertical spatial resolution [10]. AFMs do not require any special sample preparation. AFM sample preparation is relatively simple and less time consuming [5]. AFMs use cantilevers and very sharp tips fabricated using semiconductor processing techniques [5]. The AFM provides a 3D profile on a nanoscale, by measuring forces between a sharp tip (radius less than 10nm) and surface at very short distance (0.2–10 nm tip-sample separation). AFM tips attached to flexible cantilevers allow users to record small forces between the tip and the surface as the AFM tip gently touches the sample. These forces can be described using Hooke's law (Equation 1.1):

$$F = -k \cdot x \tag{1.1}$$

where F is the tip-sample force, k is the spring constant of the cantilever, and x is the extent of cantilever deflection [1, 7]. AFM systems are often equipped with optical microscopes to allow users to select locations for AFM scanning [1, 7]. AFM utilizes intermolecular forces between the tip and the surface to obtain the topographic information on the surface and other physical properties [10]. AFM systems provide height maps, generated from height measurements taken at multiple positions along the sample surface [11]. Height data appears as colors or tints, some variant of dark-is-low/bright-is-high, with a gradient of color or grayscale in between [9].

1.2.1 AFM Components

The basic components of an AFM are the tip, the cantilever, the scanner, the laser, a data processor, and a photodiode, as shown in Figure 1.2 [1, 7]. Additional components include a base, a scanner, and an optical head, in

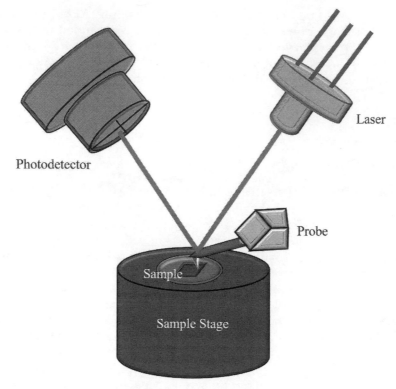

FIGURE 1.2
Diagram of atomic force microscope components and operation.

which a holder for the cantilever is mounted [12]. The optical head comprises of a special holder for the probe, as well as the optics (laser and photodiode) used in the beam-deflection detection scheme. The base contains electronic circuitry and is the interface between controller and the microscope. It also serves as physical holder for the scanner. A stepper motor, used for the coarse and fine approach between tip and sample, may be included [12]. The scanner (Figure 1.3) contains ceramic piezo elements that allow sample scanning in the x-y plane with sub-nanometer precision. The piezo elements move in z-direction to maintain a constant force between the tip and the sample [4]

The unique resolution of AFM techniques is due to ultra-sharp tips, with radii of typically 4–60 nm, attached to a flexible cantilever (Figure 1.4) [4]. The resolution of an AFM depends strongly on the shape of the tip. The smaller the tip is, the smaller the surface area sampled by the tip [4]. Pyramidal AFM tips are micro-fabricated from silicon (Si) or silicon nitride (Si_3N_4). Typical radii are 10 nm for etched Si tips and 20–60 nm for standard Si_3N_4 tips [4].

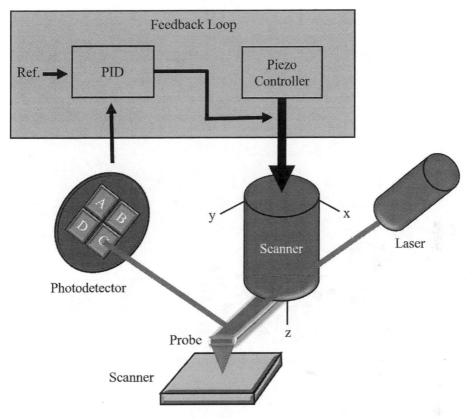

FIGURE 1.3
AFM system with piezoelectric scanner and control electronics.

1.3 Basic AFM Operation

Upward or downward cantilever deflection allows indirect measurement of heights [9]. The cantilever deflects in the z-direction due to the surface topography during tip scanning over the sample surface [4]. Cantilever bending makes tracking of vertical tip motion possible. Monitoring the vertical motion of the cantilever employs the use of a laser beam reflected off the cantilever and onto a split photodiode, the output of which gauges the position of the laser spot [9]. A four-segment photodiode detects the deflection of the cantilever through a laser beam focused on and reflected from the rear of the cantilever [4]. Additionally, monitoring lateral forces that torque the tip, causing the cantilever to twist, is possible due to the horizontal movement of the laser spot [9]. A computer processes the electrical differential signal of the photodiode

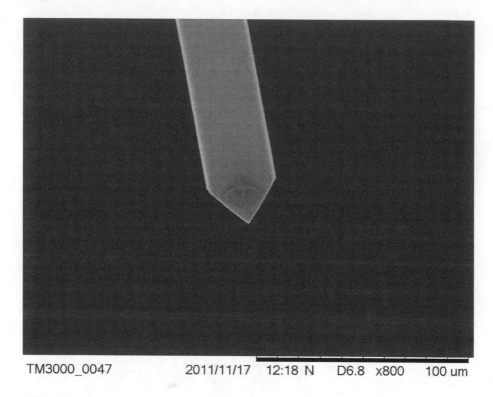

TM3000_0047   2011/11/17   12:18 N   D6.8 x800   100 um

FIGURE 1.4
Scanning electron microscope image of an AFM probe.

obtained from each point of the surface and generates a feedback signal for the scanner to maintain a constant force between the tip and sample. Processing this information results in the production of topographic images of sample surfaces [4].

1.3.1 AFM Modes

There are different modes of operation, which differ in the nature of tip-sample interaction. Interactions can be attractive or repulsive, ultimately setting the distance between the tip and the sample. In dynamic modes, such as tapping and non-contact modes, an oscillating tip generates amplitude or frequency data that serves as the feedback signal [13]. In contact mode, users bring the tip into contact with a surface and the tip scans laterally. This motion measures the vertical tip movement as the cantilever bends up and down to gauge surface height [9]. The force exerted on the tip varies with the difference in the surface height and thus leads to the bending of the cantilever [5].

1.3.2 Feedback Electronics

In static mode, also known as contact mode, the scanner moves the tip over the sample surface while maintaining a constant deflection [13]. While scanning, a feedback loop maintains the force between the tip and the sample constant by adjusting pixel by pixel the scanner height, so that the image acquisition involves plotting the height position versus its position on the sample [13]. Piezoelectric materials raise or lower the cantilever to maintain a constant bending of the cantilever [6]. As a result, the AFM records images of surface topography under a constant applied force (in the low nN range), which is optimized to produce maximal resolution without damaging the sample [11].

1.3.3 Laser Beam Detection

During imaging, a laser beam continuously reflects from the top of the cantilever toward a position-sensitive, four-quadrant photodiode. This laser beam detects the bend occurring in the cantilever and calculates the actual position of the cantilever [5]. Using this approach, a laser focused on the cantilever produces reflected light directed onto the photodiode. When the voltage from the top photodiode segment equals the voltage from the bottom photodiode segment, a null condition occurs. A small cantilever deflection disrupts this null condition, producing photodiode output proportional to deflection [3].

1.4 Forces in AFM

As the tip scans the sample surface, the force between the tip and the sample varies. Various scanning modes involve in different tip-sample forces. In the contact region, the distance between the tip of the cantilever and the sample surface is less than a few angstroms. Thus the probe experiences repulsive van der Waals forces. In the non-contact region, the tip is several tens to hundreds of angstroms away from the sample surface and hence experiences an attractive van der Waals forces. Intermittent or tapping mode fluctuates between repulsive and attractive forces [5].

1.5 AFM Applications

AFM is one of the most important techniques for the investigation of surfaces at the nanometer scale [8]. The main application of AFM is

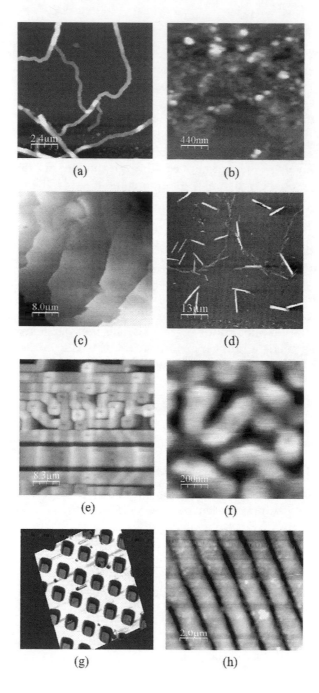

FIGURE 1.5
Topographical AFM images of silver nanowires (a), gold nanoparticles (b), hair (c), gold nanorods (d), computer chip (e), polystyrene-polymethylmethacrylate copolymer (f), calibration grating (g), and compact disc (h).

high-resolution imaging of different material surfaces including metals, polymers, ceramics, biomolecules, or cells. When ultra-flat and rigid surfaces are used, acquisition of atomic-scale resolution is possible [4]. Adding to this, AFM allows the measurement of normal and torsional deflections of the cantilever to map two-dimensional variations of physical quantities with nanoscale accuracy [14]. AFM has other noteworthy uses, including (1) the measurement of the binding force between a molecule and its ligand, (2) the determination of the elastic properties of cells, and (3) the evaluation of the friction coefficient of materials [11]. There is increasing use of AFM for force measurements and mapping of local mechanical surface properties [15]. Phase imaging allows the measurement of elasticity and sample viscosities. It is also possible to generate and collect detailed information about heterogeneities on sample surfaces with nanometer resolution [8]. Figures 1.5 a–h show examples of topographical images obtained by scanning various samples.

1.6 AFM Lithography

AFM also serves as an ultra-precise micro-manipulator, making it possible to manipulate individual atoms and molecules, yielding AFM applications that involve nanofabrication and nanomachining [6].

1.6.1 Nanoshaving

Nanoshaving involves the displacement of adsorbate molecules using an AFM tip [16]. Displacement occurs while conducting AFM scans with tip-sample forces higher than normal (Figure 1.6). Afterward, imaging of the resulting pattern involves lower tip-sample forces. It is possible to fabricate holes in trenches in molecular films such as self-assembled monolayers (SAMs) using this technique [16].

1.6.2 Nanografting

In this technique, a substrate containing a self-assembled monolayer and the AFM probe are immersed in a solution containing alkanethiol molecules that are not identical to the molecules adsorbed on the substrate surface [16]. Next, a nanoshaving step occurs in order to remove adsorbed molecules from their adsorption sites. The alkanethiol molecules in the solution are then adsorbed to the newly exposed gold surface. Afterward, reduced tip-sample forces allow imaging of the resulting nanostructures [16].

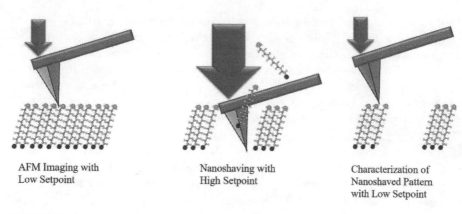

AFM Imaging with
Low Setpoint

Nanoshaving with
High Setpoint

Characterization of
Nanoshaved Pattern
with Low Setpoint

FIGURE 1.6
Nanoshaving a pattern in a self-assembled monolayer.

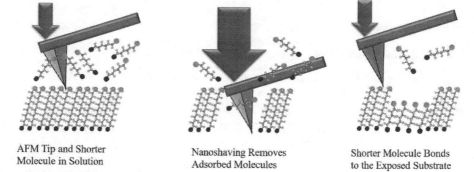

AFM Tip and Shorter
Molecule in Solution

Nanoshaving Removes
Adsorbed Molecules

Shorter Molecule Bonds
to the Exposed Substrate

FIGURE 1.7
Nanografting procedure with and AFM probe.

1.6.3 Dip Pen Nanolithography (DPN)

Dip-pen nanolithography (DPN) is a technique invented by Chad Mirkin, who first reported this technique in a scientific paper published in 1999 [17]. In this technique, materials that are initially on the tip diffuse to the surface as the tip scans the sample surface. Users can perform DPN while the AFM operates in static or dynamic mode. DPN is using the AFM tip as a nanoscale "pen" [17]. Molecular transport of the molecular "ink" depends on parameters such as composition, tip-sample contact, molecular mobility of the ink, ink solubility, temperature, and humidity. Linewidths of the resulting patterns are highly dependent on the writing speed. Studies have suggested that the molecular ink likely diffuses through a solvent meniscus condensed between the tip and the substrate [17]. However, writing experiments described in the literature have demonstrated that DPN is possible at extremely low relative humidity or with molecules that are insoluble in water. This suggests that additional tip-sample mechanisms may play a role in DPN [17]. Dip-pen nanolithography (DPN) is a lithographic technique that offers high-throughput fabrication of nanostructures [18]. This technique affords researchers the opportunity of nanoscale patterning using the AFM tip as a "pen." Since DPN relies on the compatibility between the "ink" and the substrate, this technique is limited by the nature of ink material and substrate [18]. For this reason, researchers have relied heavily on thiol-gold chemistries for nanofabrication using DPN. This method of fabrication depends on several factors, including dwell time, writing speed, solubility, and viscosity of the ink [18].

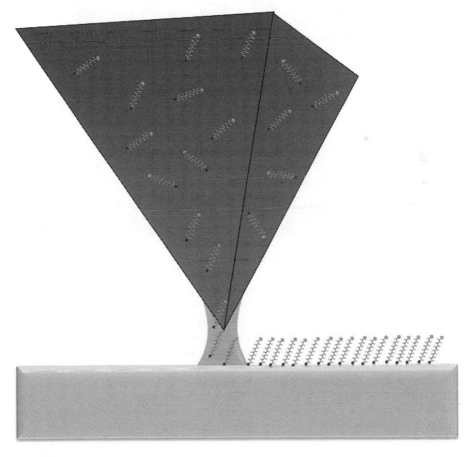

FIGURE 1.8
Dip-pen nanolithography using the AFM tip as a "pen" and an alkanethiol solution as "ink."

End-of-Chapter Questions

1. Which of the following techniques involves the measurement of current between a conductive tip and sample for nanoscale characterization?

    a. atomic force microscopy (AFM)

    b. scanning tunneling microscopy (STM)

    c. optical microscopy

    d. a and c

2. True or False. Scanning probe microscopes (SPMs) utilize magnetic lenses for high resolution imaging of nanoscale surface features.

    a. True

    b. False

3. True or False. Generally SPM probes remain stationary during imaging.

    a. True

    b. False

4. SPMs can image under the following conditions. Choose all that apply.

    a. air

    b. vacuum

    c. liquid

    d. gaseous atmospheres

5. Which of the following STM techniques is limited to conductive samples?

    a. AFM

    b. STM

    c. AFM and STM

6. Monitoring one or more interactions between the tip and the sample surface allows AFM users to determine:

    a. surface morphology

    b. surface and subsurface organization

    c. physical and chemical properties

    d. all of the above

7. AFM tip-sample distance range from _____ nm to _____ nm.

    a. 50, 100

    b. 25, 20

   c. 5, 10

   d. 0.2, 10

8. Which of the following is the correct Hooke's law equation?

   a. $F = k \times x$

   b. $F = \dfrac{k}{x}$

   c. $F = \dfrac{x}{k}$

   d. $F = -k \times x$

9. Which of the following allows users to select locations for AFM scanning?

   a. laser

   b. optical microscope

   c. photodiode

   d. cantilever

10. Which of the following allows the measurement of vertical cantilever motion? Choose all that apply.

   a. a laser beam

   b. photodiode

   c. a scanner

   d. a tunneling current

11. An oscillating tip generates amplitude or frequency data when the AFM operates in:

   a. tapping mode

   b. contact mode

   c. non-contact mode

   d. a and c

12. Vertical tip movement is used to gauge surface height in:

   a. tapping mode

   b. contact mode

   c. non-contact mode

   d. a and c

13. The cantilever maintains a constant deflection during imaging in:

   a. tapping mode

   b. contact mode

c. non-contact mode

d. a and c

14. Constant cantilever deflection is maintained by:
    Choose all that apply.

    a. raising the probe

    b. lowering the probe

    c. twisting the cantilever

    d. oscillating the cantilever at the resonance frequency

15. What type of materials move the probe to maintain constant cantilever deflection?

    a. piezoelectric materials

    b. electromechanical motors

    c. temperature-sensitive strips

    d. none of the above

16. The distance between the tip of the cantilever and the sample surface is a few angstroms during:

    a. contact mode

    b. tapping mode

    c. non-contact mode

    d. oscillating modes

17. The tip experiences repulsive van der Waals forces when the AFM operates in:

    a. contact mode

    b. tapping mode

    c. non-contact mode

    d. oscillating modes

18. The tip is several tens to hundreds of angstroms away from the sample surface during:

    a. contact mode

    b. non-contact mode

    c. tapping mode

    d. oscillating modes

19. The tip experiences attractive van der Waals forces when the AFM operates in:

    a. contact mode

    b. tapping mode

    c. oscillating modes

    d. non-contact mode

20. Which mode fluctuates between repulsive and attractive forces?

    a. non-contact mode

    b. oscillating modes

    c. contact mode

    d. tapping mode

21. What AFM component maintains a constant cantilever deflection?

    a. laser

    b. photodiode

    c. probe

    d. feedback loop

22. Constant cantilever deflection is achieved by adjusting:

    a. laser intensity

    b. vibration amplitude

    c. scanner height

    d. photodetector sensitivity

23. What type of samples are required for atomic-scale resolution?
Choose all that apply.

    a. biological

    b. polymers

    c. rigid

    d. flat

24. True or False. AFM can only monitor normal deflection of the cantilever.

    a. True

    b. False

25. True or False. AFM systems cannot perform nanomechanical measurements on samples.

    a. True

    b. False

26. Which of the following techniques requires the use of high tip-sample forces for nanofabrication?
Choose all that apply.

    a. DPN

    b. nanoshaving

    c. nanografting

References

[1] R. R. De Oliveira, D. A. Albuquerque, T. G. Cruz, F. M. Yamaji and F. L. Leite, "Measurement of the nanoscale roughness by atomic force microscopy: Basic principles and applications," in *Atomic Force MIcroscopy—Imaging, Measuring, and Mainpulating Surfaces at the Atomic Scale*, V. Bellitto, Ed., Rijeka, InTech, 2012, pp. 147–174.

[2] A. S. Foster, W. A. Hofer and A. L. Shluger, "Quantitative modelling in scanning probe microscopy," *Curr. Opin. Solid St. M.*, vol. 5, pp. 427–434, 2001.

[3] R. Reifenberger, "Introduction to Scanning Probe Microscopy," *Fundamentals of Atomic Force Microscopy—Part I: Foundations*, Hackensack: World Scientific Publishing Co., 2016, pp. 1–20.

[4] K. D. Jandt, "Atomic force microscopy of biomaterials surfaces and interfaces," *Surf. Sci.*, vol. 491, pp. 303–332, 2001.

[5] S. Chatterjee, S. S. Gadad and T. K. Kundu, "Atomic force microscopy— A tool to unveil the mystery of biological systems," *Resonance*, pp. 622–642, July 2010.

[6] B. Bhushan and O. Marti, "Scanning probe microscopy—Principle of operation, instrumentation, and probes," in *Nanotribology and Nanomechanics*, B. Bhushan, Ed., Berlin, Springer-Verlag, 2005, pp. 41–115.

[7] P. Eaton and P. West, *Atomic Force Microscopy*, Oxford: Oxford University Press, 2010.

[8] M. Marrese, V. Guarino and L. Ambrosio, "Atomic force microscopy: A powerful tool to address scaffold design in tissue engineering," *J. Funct. Biomater.*, vol. 8, pp. 1–20, 2017.

[9] G. Haugstad, *Atomic Force Microscopy—Understanding Basic Modes and Advanced Applications*, Hoboken: John Wiley and Sons, Inc., 2012.

[10] V. V. Tsukruk and S. Singamaneni, *Scanning Probe Microscopy of Soft Matter: Fundamentals and Practices*, Weinheim: Wiley-VCH Verlag GmbH & Co. KGaA., 2012.

[11] P. P. Lehenkari, G. T. Charras, S. A. Nesbitt and M. A. Horton, "New technologies in scanning probe microscopy for studying molecular interactions in cells," *Expert Rev. Mol. Med.*, pp. 1–19, 8 March 2000.

[12] "Atomic force microscopy in practice," in *Scanning Force Microscopy of Polymers*, Berlin, Springer-Verlag, 2010, pp. 25–27.

[13] C. Musumeci, "Advanced scanning probe microscopy of graphene and other 2D materials," *Crystals*, vol. 7, pp. 1–19, 2017.

[14] R. Buzio and U. Valbusa, "Nanolubrication studied by contact-mode atomic force microscopy," in *Modern Research and Educational Topics in Microscopy*, A. Mendez-Vilas and J. Diaz, Eds., Badajoz, Formatex, 2007, pp. 491–499.

[15] A. F. Sarioglu and O. Solgaard, "Time-resolved tapping-mode atomic force microscopy," in *Scanning Probe Microscopy in Nanoscience and Nanotechnology*, B. Bhushan, Ed., Berlin, Springer-Verlag, 2011, pp. 3–37.

[16] G. Yang, N. A. Amro and G. Liu, "Scanning probe lithography of self-assembled monolayers," in *Proceedings of SPIE*, Bellingham, 2003.

[17] L. G. Rosa and J. Liang, "Atomic force microscopy nanolithography: Dip-pen, nanoshaving, nanografting, tapping mode, electrochemical and thermal nanolithography," *J. Phys. Condens. Matter*, vol. 21, pp. 1–18, 2009.

[18] S. Sharma, A. Salehi-Reyhani, A. Bahrami, E. Intisar, H. Santhanam, K. Michelakis and A. Cass, "Method for fabricating nanostructures via nanotemplates using dip-pen nanolithography," *Micro Nano Lett.*, vol. 7, pp. 1038–1040, 2012.

2

Tip-Sample Forces

2.0 Key Objectives

- Become familiar with the following tip-sample forces:
  - Van der Waals forces
  - Keesom forces
  - Debye forces
  - Repulsive forces
  - Capillary forces
- Learn how to interpret force curve diagrams.

2.1 Introduction

Tip-sample forces are used to image surfaces. These forces provide surface height information and data regarding material properties, interfaces, and even molecules. Measuring tip-sample forces across a sample surface allows determination of spatial distribution of materials [1]. The cantilever bends vertically because of repulsive (upward bend) or attractive (downward bend) interactions, as shown in Figures 2.1a and 2.1b, respectively. The magnitude of the forces acting between the tip and sample surface varies, depending on the sample nature, imaging mode, and conditions used in the measurements [2].

The AFM measures forces as small as 1 nN between the AFM tip and sample [3]. Tip-sample, distance-dependent interactions can be repulsive, van der Waals, electrostatic, or magnetic in nature [4], as shown in Figure 2.2.

Most of the forces exerted between the tip and sample surface relate to dipoles. The atoms or molecules in the tip and sample have temporary dipoles due to an internal displacement of charges. When two inert, spherical atoms with fluctuating charge distributions are in close proximity, the electrons of one atom tend to repel the nearest electrons of the other atom. As a result, a dipole on one atom induces a temporary dipole on the adjacent

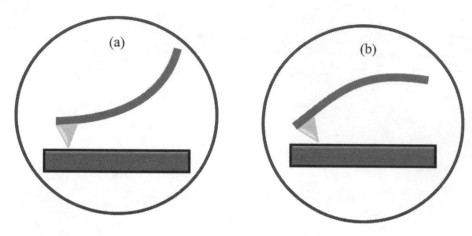

FIGURE 2.1
The upward bend (a) and downward bend (b) of a cantilever due to tip-sample interactions.

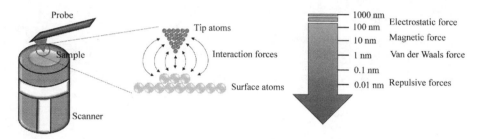

FIGURE 2.2
Interactive forces found between the AFM tip and sample surfaces at various distances.

atom, resulting in attraction [1]. Tip-sample interactions can be graphically represented using the Lennard-Jones Potential Energy Diagram (Figure 2.3). This diagram shows that the potential energy of attraction between atoms increases with the inverse distance to the 6th power. Additionally, the potential energy of repulsion between atoms increases with the inverse distance to the 12th power [1].

2.2 Van der Waals Forces

Van der Waals forces are present in all AFM operating modes. Fluctuations in the electron cloud of atoms is responsible for these forces [4]. The van der Waals force exerted between atoms and/or molecules is inclusive of three

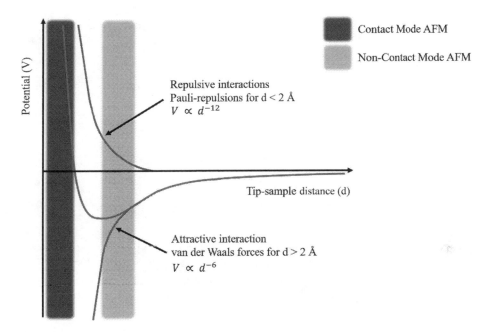

FIGURE 2.3
Lenard-Jones potential energy diagram depicting the potential energy versus distance relationship for attractive and repulsive forces.

forces, all proportional to $1/r^6$, where r indicates the distance between the atoms or molecules. The constituent forces include Keesom forces, $w_{K(r)}$; Debye forces, $w_{D(r)}$; and London dispersion forces, $w_{L(r)}$ [5]. London dispersion forces are the most important component because these forces exist between all molecules or atoms. London dispersion forces are felt between molecules with temporary dipoles calculated using Equation 2.1:

$$w_{L(r)} = -\frac{3}{2} \frac{\alpha_{02}\alpha_{01}}{(4\pi\varepsilon_0)^2 r^6} \frac{h\nu_1\nu_2}{\nu_1 + \nu_2} \tag{2.1}$$

The terms $h\nu_1$ and $h\nu_2$ represent the first ionization potentials of the molecules and h is Planck's constant [5]. Keesom forces involve interactions between permanent dipoles, quantified using Equation 2.2:

$$w_{K(r)} = -\frac{u_1^2 u_2^2}{3(4\pi\varepsilon_0 \varepsilon)^2 k_B T r^6} \tag{2.2}$$

Dipole moments of the molecules are represented by the terms u_1 and u_2, ε the dielectric constant of the medium, k_B is Boltzmann's constant, and T is the temperature [5]. Lastly, Debye forces arise from dipole-induced dipole interactions between two atoms or molecules (Equation 2.3):

$$w_{D(r)} = -\frac{u_1^2 \alpha_{02} + u_2^2 \alpha_{01}}{(4\pi\varepsilon_0\varepsilon)^2 r^6} \tag{2.3}$$

Here, α_{01} and α_{02} represent the electronic polarizabilities of the molecules [5]. The total tip-sample van der Waals interaction, inclusive of all three constituents, is the sum of the three terms as shown in Equation 2.4:

$$w_{vdW} = -\frac{3k_B T}{(4\pi\varepsilon_0)^2 r^6}\left(\frac{u_1^2}{3k_B T} + \alpha_{01}\right)\left(\frac{u_2^2}{3k_B T} + \alpha_{02}\right) - \frac{3}{2}\frac{\alpha_{01}\alpha_{01}}{(4\pi\varepsilon_0)^2 r^6}\frac{h\nu_1\nu_2}{\nu_1 + \nu_2} \tag{2.4}$$

The first term shown in Equation 2.4 represents the Keesom and Debye energies. This term is specific for interactions between polar molecules. The second term quantifies the London dispersion forces acting between every molecule. It is also important to note that this term is known as the dispersion contribution [5].

2.3 Repulsive Forces

Repulsive forces occur between the AFM tip and sample surface due to electrostatic repulsions between atoms or molecules close enough for electron orbitals to overlap [2]. The Pauli exclusion principle, which prevents electrons from occupying states with the same quantum numbers, governs these repulsive forces. Coulombic repulsion between nuclei contribute to these forces as well [6]. When repulsive forces predominate, the tip and sample are in contact [2].

2.4 Capillary Forces

Under ambient conditions, moisture can condense on the tip, producing capillary forces [7]. Either the surface contains a monolayer of water, or condensation of liquid between the AFM tip and sample occurs. In each case, a liquid meniscus strongly pulls the tip toward the sample surface (Figure 2.4) [7].

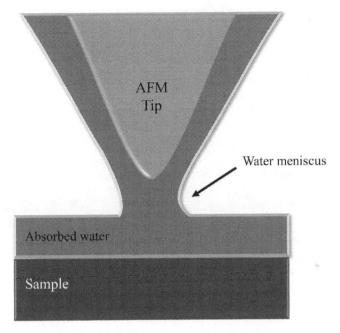

FIGURE 2.4
A liquid meniscus forming between an AFM tip and sample surface.

It is possible for capillary forces to cover the contributions of other forces. Users can circumvent capillary forces by imaging in a dry, N_2, or Ar atmosphere, or in liquid environments [7]. The Laplace pressure is the dominant interaction in capillary forces and is determined using Equation 2.5:

$$P = \gamma \left(\frac{1}{r_1} + \frac{1}{r_2} \right) = \frac{\gamma}{r_k} \tag{2.5}$$

where γ is the surface tension, r_1 and r_2 are the radii of the meniscus, and r_k is the Kelvin radius. Determination of the magnitude of the capillary force involves Equation 2.6:

$$F_{capillary} = \frac{\gamma A}{r_k} = \frac{\gamma 2\pi R d}{r_k} \tag{2.6}$$

where A is the contact area of the meniscus, R is the tip radius, and d is the tip penetration depth into thin molecular film [6].

2.5 Force Curves

In addition to recording surface topography and elucidating the chemical properties of samples, AFMs can provide quantitative information of forces between the tip and the sample as a function of tip-sample distance in the form of a force curve [8]. Force curve production involves moving the tip toward the sample in a normal direction while recording the vertical position of the tip and the deflection of the cantilever [7]. A typical force curve is shown in Figure 2.5. At the first stage of the cycle (1), a large distance separates the tip and the sample, with no interaction between the two. As the tip-sample distance decreases, long- and short-range attractive forces pull on the tip. Once these forces acting on the tip exceeds cantilever stiffness, it jumps into contact with the sample surface, represented by point (2) [2]. At point (3), the tip and sample are in contact and cantilever deflection is due to repulsions between the orbitals of the tip and sample atoms. It is noteworthy to mention that region (1–2) is the approach curve. In region (3–4), which is generally linear, the elastic properties of the sample can be determined. The slope of the curve in the contact region is a function of the

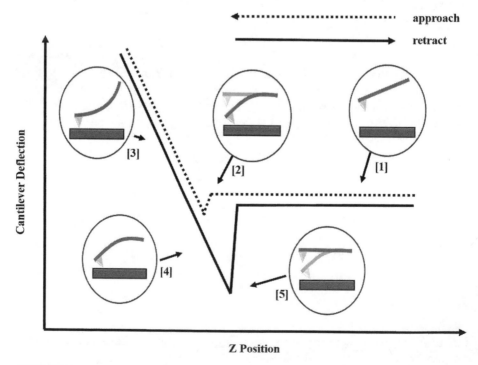

FIGURE 2.5
Force curve showing tip behavior a various tip-sample distances.

elastic modulus of the tip and sample. The slope allows users to infer the sample hardness or indicate sample response at different loads. The segment (3–5) represents the withdrawal curve. During withdrawal, adhesion or bonds formed during contact with the surface cause the tip to adhere to the sample (4). As the tip continues retracting, the spring constant of the cantilever overcomes the adhesion forces, and the cantilever pulls off sharply, springing upward to its non-deflected position (5). The jump-off contact occurs when the cantilever spring constant is greater than the tip-sample adhesive forces (pull-off force) [2].

2.5.1 Force Spectroscopy

Force spectroscopy involves using force curve data to determine mechanical properties such as elastic modulus and breaking strength. The maximum tension that a material can withstand and tensile strength can also be determined using force spectroscopy [9]. Upon contact with the sample surface, the AFM system monitors forces the tip experiences while cantilever deflection changes. Measuring forces as a function of tip-sample separation allows investigation of several characteristics of the tip, sample, or medium in-between [7].

2.6 Laboratory Exercise: Force Curve Analysis of Metallized Polymer Patterns

2.6.1 Laboratory Objectives

The objectives of this experiment include:

- microfabrication of polymeric grids
- acquisition of force curves
- determination of mechanical properties from force curves

2.6.2 Materials and Procedures

The recommended samples for use in this experiment are silver nanogrids produced from the metallization of PVP templates using dilute concentrations of silver nitrate and sodium citrate as a reducing agent. This silver nanogrid fabrication is reported in the literature [10]. A 2 cm imaging disk and double-sided tape is needed to mount the sample. The AFM software used in this lab should have a force spectroscopy option. Use a diamond scribe to cut a 2 cm × 2 cm glass square from a clean microscope slide. PVP

TABLE 2.1

Recommended Parameters for Force Curve Acquisition

Parameter	Value
Setpoint	13 nN
Start Offset	−1 μm
Range	1 μm
Modulation Time	1 s
Data Points	256

grids are microcontact printed on the glass squares following the procedures in the literature [11]. Fabricate silver nanogrids on glass following procedures stated in the literature [10]. The glass square containing the PVP grid and the silver nanogrid can then be mounted on an imaging disc using double-sided tape. Set up the AFM for contact mode imaging according to the AFM manual. Use the force spectroscopy option to determine force curves for the PVP template and the PVP template metallized with silver. Table 2.1 shows the recommended parameters for the force curve analyses.

2.6.3 Sample Data and Results

Topographical images of bare PVP and metallized polymer grids are show in Figures 2.6a and 2.7a respectively. Figures 2.6b and 2.7b show force curve analyses of bare PVP and metallized grids respectively. Markings are included for slope and pull off force determination. Slopes allow determination of the spring constant (k) of the sample using Equation 2.7 [12]. Mechanical data obtained from the force constant data is in Table 2.2.

$$\frac{1}{slope} = \frac{1}{k_{cantilever}} + \frac{1}{k_{sample}} \tag{2.7}$$

Post-Lab Questions

1. According to the data shown in Table 2.2, which of the following samples has the highest slope?
   a. PVP grid
   b. silver nanogrid

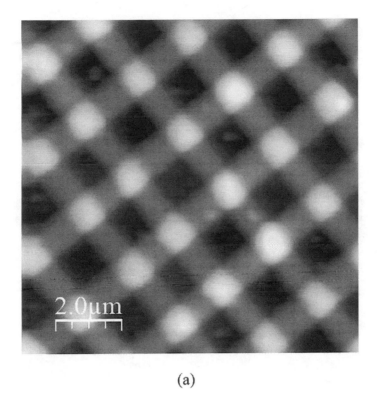

(a)

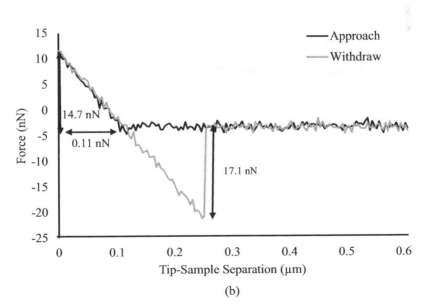

(b)

FIGURE 2.6
Contact mode image (a) and force curve analysis (b) of PVP grid.

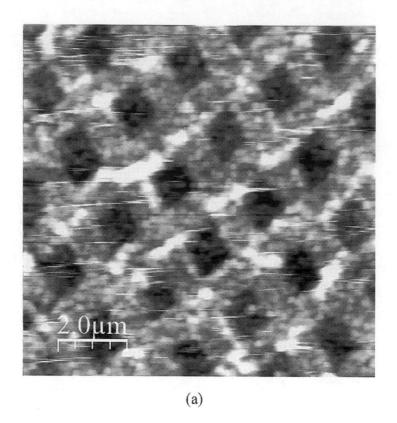

(a)

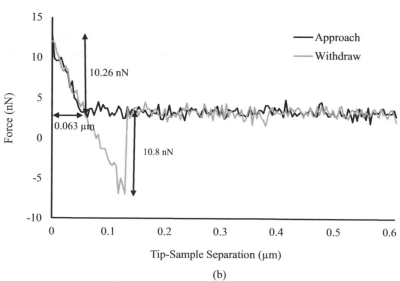

(b)

FIGURE 2.7
Contact mode image (a) and force curve analysis (b) of silver nanogrid.

TABLE 2.2

Spring Constant Data for PVP and Silver Nanogrids

Parameter	PVP Grid	Silver Nanogrid
Slope	0.13 N/m	0.163 N/m
Spring Constant	0.37 N/m	0.88 N/m
Pull-Off Force	17.1 nN	10.8 nN

2. According to the data shown in Table 2.2, which of the following samples has the highest spring constant?

   a. PVP grid

   b. silver nanogrid

3. Which of the following forces is likely to exist between the tip and the sample with the highest spring constant?

   a. attractive

   b. repulsive

4. Which sample has the highest pull-off force?

   a. PVP grid

   b. silver nanogrid

5. Based on the pull-off force data, greater adhesive forces are observed between the tip and the:

   a. PVP grid

   b. silver nanogrid

6. Which of the following forces are likely to exist between the tip and sample with the largest pull-off force?

   a. Keesom forces

   b. Debye forces

   c. London dispersion forces

   d. Capillary forces

7. Pull-off force data is obtained from the _____ curve.

   a. approach

   b. withdraw

8. Spring constant data is obtained from the _____ curve.

   a. approach

   b. withdraw

9. Snap into contact occurs during:

a. approach

b. withdraw

End-of-Chapter Questions

1. An upward cantilever bend is due to _____ forces, and a downward cantilever bend is due to _____ forces.

a. attractive, repulsive

b. repulsive, attractive

2. AFM systems can measure forces as small as _____ nN between the AFM tip and sample.

a. 100

b. 75

c. 25

d. 1

3. Repulsive forces are present when tip-sample distances are approximately _____, and attractive van der Waals forces are present when tip-sample distances are approximately _____.

a. 1000 nm, 1 nm

b. 0.01 nm, 1 nm

c. 1 nm, 100 nm

d. 1 nm, 0.1 nm

4. The atoms or molecules found in the tip and sample have:

a. temporary dipoles

b. permanent dipoles

5. The dipoles found in the tip and sample are formed using:

a. stationary charge distributions

b. fluctuating charge distributions

6. Given that r represents the distance between atoms or molecules, the magnitude of van der Waals forces acting between them is proportional to:

a. $\dfrac{1}{r^{12}}$

b. r^{12}

c. r^6

d. $\dfrac{1}{r^6}$

7. Given that r represents the distance between atoms or molecules, the magnitude of repulsive forces acting between them is proportional to:

a. $\dfrac{1}{r^{12}}$

b. r^{12}

c. r^6

d. $\dfrac{1}{r^6}$

8. Which of the following forces act between the tip and sample during AFM imaging?
Choose all that apply.

a. Keesom forces

b. Debye forces

c. London dispersion forces

9. Which of the following is the most prominent force acting between the tip and sample during AFM imaging?

a. Keesom forces

b. Debye forces

c. London dispersion forces

10. Which of the following forces exist between molecules with temporary dipoles?

a. Keesom forces

b. Debye forces

c. London dispersion forces

11. Which of the following forces exist between molecules with permanent dipoles?

a. Keesom forces

b. Debye forces

c. London dispersion forces

12. Which of the following forces originate from dipole-induced dipole interactions?

    a. Keesom forces

    b. Debye forces

    c. London dispersion forces

13. Electrostatic interactions between atoms are molecules close enough for orbitals to overlap result in the formation of:

    a. Keesom forces

    b. Debye forces

    c. London dispersion forces

    d. Repulsive forces

14. Which of the following forces is produced from condensation forming between the tip and sample while imaging under ambient conditions?

    a. London dispersion forces

    b. Debye forces

    c. Capillary forces

    d. Repulsive force

15. Which of the following reduces the effect of capillary forces? *Choose all that apply.*

    a. low-humidity environments

    c. nitrogen environments

    c. argon environments

    d. liquid environments

16. Which of the following is the dominant interaction in capillary forces?

    a. van der Waals interactions

    b. Laplace Pressure

    c. Keesom forces

    d. Debye forces

17. Force curves are generated by moving the tip in a _____ direction.

    a. normal

    b. lateral

18. During a force curve experiment, the extent that the cantilever _____ is recorded.

    a. twists

    b. deflects

19. In region 1 of the force curve shown in Figure 2.5, which of the following forces is present between the tip and sample?
    a. attractive
    b. repulsive
    c. attractive and repulsive
    d. none of the above

20. Consider the force curve shown in Figure 2.5. Attractive forces exist between the tip and the sample in which of the following regions? *Choose all that apply.*
    a. 1
    b. 2
    c. 3
    d. 4
    e. 5

21. Consider the force curve shown in Figure 2.5. Repulsive forces exist between the tip and the sample in which of the following regions?
    a. 1
    b. 2
    c. 3
    d. 4
    e. 5

22. Consider the force curve shown in Figure 2.5. Which of the following regions is likely to provide information regarding adhesive properties (pull-off force)?
    a. 1
    b. 2
    c. 3
    d. 4
    e. 5

23. Consider the force curve shown in Figure 2.5. Which of the following regions is likely to provide information regarding the elastic properties of samples? *Choose all that apply.*
    a. 1
    b. 2
    c. 3
    d. 4
    e. 5

24. Consider the force curve shown in Figure 2.5. It is possible to acquire modulus data by taking the slope of region(s):
    Choose all that apply.

    a. 1
    b. 2
    c. 3
    d. 4
    e. 5

25. Elastic data is determined by analyzing the _____ curve.

    a. approach
    b. withdrawal

26. Adhesion data is determined by analyzing the _____ curve.

    a. approach
    b. withdrawal

27. True or False. Force spectroscopy can only determine breaking strength of nanomaterials.

    a. True
    b. False

28. Force spectroscopy can determine mechanical properties of:

    a. the tip only
    b. the sample only
    c. the tip and the sample

References

[1] G. Haugstad, *Atomic Force Microscopy—Understanding Basic Modes and Advanced Applications*, Hoboken: John Wiley and Sons, Inc., 2012.
[2] F. L. Leite, L. H. C. Mattoso, O. N. Oliveira Jr. and P. S. P. Herrmann Jr., "The atomic force spectroscopy as a tool to investigate sruface forces: Basic principles and applications," in *Modern Research and Education Topics in MIcroscopy*, A. Mendez-Vilas and J. Diaz, Eds., Badajoz, Formatex, 2007.
[3] B. Bhushan and O. Marti, "Scanning probe microscopy—Principle of operation, instrumentation, and probes," in *Nanotribology and Nanomechanics*, B. Bhushan, Ed., Berlin, Springer-Verlag, 2005, pp. 41–115.
[4] F. J. Giessibl and C. F. Quate, "Exploring the nanoworld with atomic force microscopy," *Phys. Today*, vol. 12, pp. 44–50, 2006.

[5] H. J. Butt, B. Cappella and M. Kappl, "Force measurements with the atomic force microscope: Technique, interpretation and applications," *Surf. Sci. Rep.*, vol. 59, pp. 1–152, 2005.

[6] E. Meyer, "Atomic force microscopy," *Prog. Surf. Sci.*, vol. 41, pp. 3–49, 1992.

[7] U. Maver, T. Maver, Z. Peršin, M. Mozetič, A. Vesel, M. Gaberšček and K. Stana-Kleinschek, "Polymer characterization with the atomic force microscope," in *Polymer Science*, F. Yilmaz, Ed., London, InTech, 2013, pp. 113–132.

[8] K. D. Jandt, "Atomic force microscopy of biomaterials surfaces and interfaces," *Surf. Sci.*, vol. 491, pp. 303–332, 2001.

[9] C. Musumeci, "Advanced scanning probe microscopy of graphene and other 2D materials," *Crystals*, vol. 7, pp. 1–19, 2017.

[10] W. C. Sanders, R. Valcarce, P. Iles, J. S. Smith, G. Glass, J. Gomez, G. Johnson, D. Johnston, M. Morham, E. Beefus, A. Oz and M. Tomaraei, "Printing silver nanogrids on glass," *J. Chem. Educ.*, vol. 94, pp. 758–763, 2017.

[11] W. C. Sanders, "Fabrication of polyvinylpyrrolidone micro-/nanostructures utilizing microcontact printing," *J. Chem. Educ.*, vol. 92, pp. 1908–1912, 2015.

[12] M. A. Ferguson and J. J. Kozlowski, "Using AFM force curves to explore properties of elastomers," *J. Chem. Educ.*, vol. 90, pp. 364–367, 2012.

3

AFM Electronics

3.0 Key Objectives

- Become familiar with the roles of analog and digital electronics.
- Understand the roles of the photodiode, scanner, piezotubes, and motors.
- Become familiar with routine AFM parameters optimized during imaging.
- Understand the role and operation of the feedback loop.
- Understand the role of gains and how to optimize gains.

3.1 Analog and Digital Electronics

AFM electronics are responsible for x-y piezo scanner control, generating the control signal for z scanner control, x-y-z stepper motor control, production of signals for probe oscillation when tapping or noncontact modes are used, and collection of signals for computer processing [1]. Computers equipped with dual monitors allow users to view AFM images and to control parameters simultaneously. AFM operation also includes the use of a digital signal processor (DSP), shown in Figure 3.1. DSPs contain analog-to-digital converters (ADC) and digital-to-analog converters (DAC) for the processing of the signals between the scanner head and the DSP [1]. The DSP also receives analog voltage signals from the photodiode [2]. DSP chips perform necessary feedback control calculations and facilitates the x-y raster scan functions [1].

AFM electronics must be capable of amplifying small signals from the computer to hundreds of volts in order to move the piezo tubes [2]. The piezoelectric scanner is driven by a high voltage amplifier. Operational amplifiers and comparator circuits monitor voltages from the photodiode in order to apply the necessary signals to x, y, and z piezos, as shown in Figure 3.2 [1]. All computer-based systems require proprietary software to operate AFM instrumentation. Processing raw AFM images requires the use of image-processing software available free on the internet [2].

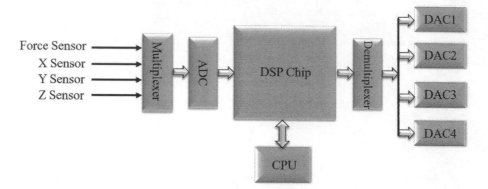

FIGURE 3.1
Analog to digital electronics in AFM systems including the DSP chip.

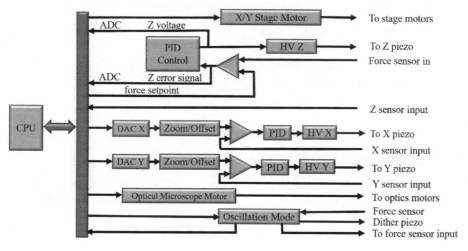

FIGURE 3.2
Block diagram of the electronics used to control the piezoelectric tubes.

3.2 AFM Components

A standard AFM system consists of a laser, a photodiode, positioning mechanisms for aligning the laser beam on the reflective side of the cantilever, and small electronics for processing the signals coming from the photodiode [2].

3.2.1 Photodiode

The photodiode is a semiconductor device that converts light striking the device into an analog voltage signal. The photodiode has four sections that allow tracking of vertical and lateral movements of the cantilever as shown in Figure 3.3. The photodiode produces a signal from the position of the reflected beam.

Photodiodes monitor the position of a laser beam reflecting from the cantilever (Figure 3.4). Vertical movement of the cantilever is measured as the difference in voltage between upper and lower quadrants of the photodiode [2].

Monitoring torsional motion involves recording the difference in voltage between the left and right quadrants of the photodiode. The torsional movement of the cantilever arises from frictional forces between the tip and the sample during scanning [2]. Analog V_A, V_B, V_C, and V_D signals produced by the detector allow determination of the vertical (Equation 3.1) and lateral (Equation 3.2) deflection signals measured in volts:

$$[V] - \frac{(V_A + V_B) - (V_C + V_D)}{V_A + V_B + V_C + V_D} \tag{3.1}$$

$$[V] = \frac{(V_A + V_D) - (V_B + V_C)}{V_A + V_B + V_C + V_D} \tag{3.2}$$

Analog signals are subsequently amplified, digitalized, and sent to a computer [3].

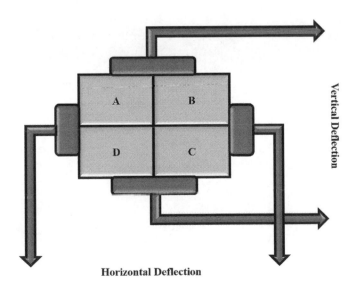

FIGURE 3.3
Four quandrant photodiode.

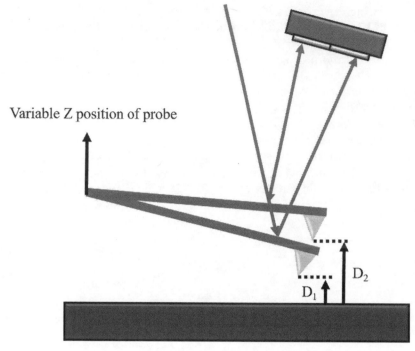

Variable Z position of probe

D_2

D_1

FIGURE 3.4
Changing position of the laser beam on the photodiode due to cantilever deflection.

3.2.2 Scanner

The scanner converts electric signals supplied from the AFM control electronics into mechanical scanning motion in order to move the tip across the sample surface [4]. While the scanner is moving across a scan line, image acquisition occurs at equally spaced intervals. The spacing between the data points is the step size. Full scan size and the number of data points per line quantifies step size. Scan sizes range from tens of angstroms to over 100 microns, with data points ranging from 64 to 512 data points per line. Additionally, 1024 data points per line are available with some systems [5].

3.2.2.1 Scanner Materials

Ceramic materials found in AFM scanners demonstrate the piezoelectric effect. This effect occurs when electrical potential appears across crystal faces when the ceramics are under pressure. This effect also occurs when the ceramics distort when an electrical field is applied. Piezoelectric materials are electromechanical transducers that convert electrical potential into

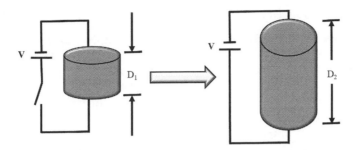

FIGURE 3.5
The expansion and contraction of piezoelectric materials when voltage is applied.

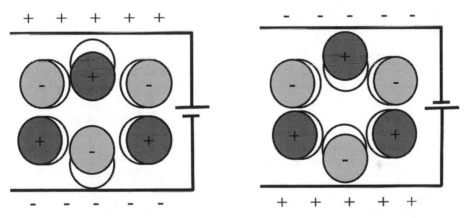

FIGURE 3.6
The displacement of ions in piezoelectric materials under the application of a voltage.

mechanical motion. Piezoelectric materials expand and contract in response to electric signals (Figure 3.5) [6].

The shape change experienced by ceramic piezoelectric materials when voltage is applied is due to the displacement of ions in the crystal lattice structure. When an electrical field is applied, the ions are displaced by electrostatic forces, resulting in the mechanical deformation of the crystal (Figure 3.6) [6].

Piezoelectric materials in AFM scanners contain lead barium titanate ($PdBaTiO_3$) or lead zirconate titanate ($PbZrTiO_3$), commonly abbreviated as PZT [1, 4]. Piezoelectric materials are useful for AFM due to their ability to control tiny movements with high accuracy. Typically, the expansion coefficient for a single piezoelectric device is 0.1 nm per volt [1].

Piezoelectric materials allow the positioning of the tip and sample with sub nanometer precision [2].

3.2.2.2 Piezotube Geometry

Piezoelectric scanners are often tube shaped (Figure 3.7a) [4]. The piezos have a tube construction consisting of a number of segments, which enables a two-dimensional scanning motion in an x-y plane as well as a motion in z-direction to vary the distance between the probe and the sample [4]. The piezotubes used in the scanners have electrodes on the inside and outside. This tube configuration provides a lot of motion and is very rigid [1]. The tubes consist of polarized, piezoelectric materials [6]. The external face of the tube is divided into four longitudinal segments of equal size and electrodes are welded to the internal and external faces of the tube (Figure 3.6b). Application of a bias voltage between the inner and all the outer electrodes encourages extension or contraction, which allows scan movement [2].

3.2.2.3 Scanner Nonlinearities

Ideally, piezoelectric materials expand and contract in direct proportion to the applied voltage. However, this is not the case; all piezoelectric materials show two primary non-ideal behaviors, hysteresis and creep [1]. Hysteresis, derived from the word *history*, is the piezoelectric material's attempt to

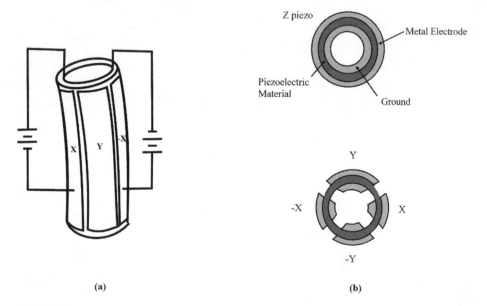

(a)   (b)

FIGURE 3.7

A schematic diagram of the piezotubes used in AFM scanners (a). The segmented structure of piezotubes (b).

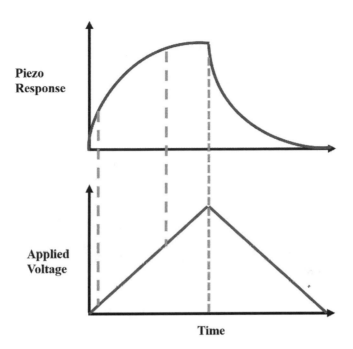

Piezo
Response

Applied
Voltage

Time

FIGURE 3.8
The nonlinear response of piezoelectric materials.

maintain its previous shape when a voltage is applied (Figure 3.8). Hysteresis produces a "bending" distortion in images, unless corrected [4]. If the applied voltage suddenly changes, the piezo-scanner's response is not all at once. It moves the majority of the distance quickly; the last part of the movement is slower as shown in Figure 3.9. When this occurs, the slow movement will produce a distortion known as creep [1]. Repeating the scan in order to allow the piezo to relax generally solves the problem [2].

3.2.3 Motors

AFM systems have stepper motors used for x-y sample translation, coarse z motion control, for approach, and zoom/focus of the optical microscope. Coarse Z motor control brings the tip and sample together from initial distances of 10s or 100s μm apart; this action is the coarse approach. After coarse approach, users should initiate the fine approach. During the fine approach, the AFM system lowers the z piezo in small increments until the tip makes contact with the sample surface. As soon as the tip makes contact with the sample surface, the feedback loop activates [1].

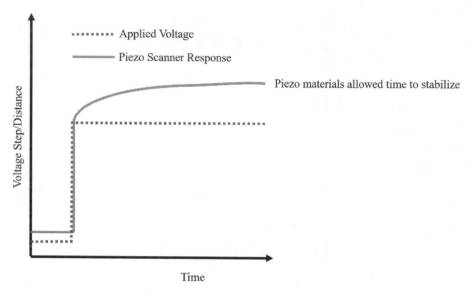

FIGURE 3.9
The onset of creep after an applied voltage.

3.3 Feedback Loop

The feedback loop controls the force between the AFM tip and sample surface. The feedback loop takes input from the photodiode and compares the signal to a setpoint value. The feedback loop keeps the tip-sample forces constant by controlling the expansion of the z piezo in the scanner (Figure 3.10) [1]. The controller compares the photodiode signal to the user-defined setpoint, and produces an error signal [7]. The error signal is then sent through a feedback controller, which in turn drives the z piezo [1]. If tip-sample forces increase, the feedback loop moves the tip away from the surface. Conversely, if tip-sample forces decrease, the probe moves toward the surface [1]. The amount the z piezo moves up and down to maintain the tip-sample distance fixed is equivalent to the sample topography [1].

3.3.1 Proportional and Integral Gains

Management of feedback control involves a proportional-integral-derivative (PID) controller, which requires selection of appropriate P, I, and D gains to track the sample surface with accuracy [1]. A PID feedback loop consists of three correcting terms, whose sum constitutes the Z-scanner position at a given time Zt following Equation 3.3 [2]:

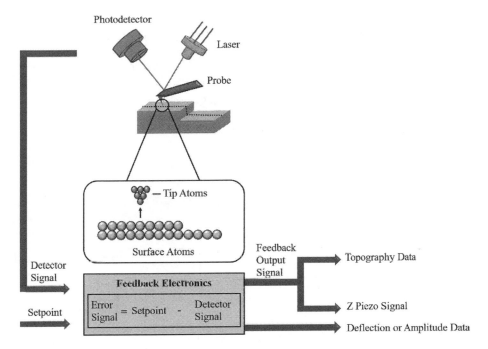

FIGURE 3.10
The feedback loop.

$$Z_t = P_{\varepsilon t} + Idt\sum_{i=0}^{t}\varepsilon_i + D\frac{\varepsilon_t - \varepsilon_{t-1}}{dt} \tag{3.3}$$

where εt is the error signal at time t (i.e., the difference between the photodiode signal at time t and the setpoint value). P, I, and D represent the proportional, integral, and derivative gains, respectively [2]. Most AFM systems only use P and I user inputs; the D term is not used. Derivative gains tend to introduce instabilities. The proportional (P) and integral (I) gains determine how quickly the feedback electronics respond to tip deflection [1]. Integral gain governs probe movement over larger features on the sample surface. Proportional gain controls the ability of the probe to track smaller, more frequent features on the sample surface [1]. After successful tip-sample engagement, the gains must be adjusted [8]. Proper feedback control requires fine-tuning the PI parameters because gains regulate the sensitivity of the feedback loop [2, 8]. PI gains govern the accuracy with which the controller maintains the constant force value. The most effective method for adjusting PI gains involves analyzing the height and deflection signals in the scope mode. The trace and retrace signals visible in the scope mode correspond to signals for individual line scans [8]. The scope mode can serve as an effective tool to optimize gains by observing the forward and backward scan signals and aiming to make them overlap [2]. Very low gains result in the feedback loop inadequately maintaining the setpoint value, and thus the "height" image does not correspond to sample topography [8]. For low gain values, the response of the system is very slow, and the backward and forward profiles are very different. As one increases gains, the response is faster, but if these parameters are too high, the system becomes unstable and the z piezo oscillates at a high frequency [2]. If the gains are set too high, the piezo scanner will show uncontrolled feedback seen as high frequency oscillations in the scope mode. Unwanted feedback can also be audible [8]. In general, the working point is in-between these two situations, and it is found using a trial-and error-procedure [2].

3.4 AFM Parameters

The setpoint is the signal that is maintained at a constant value during imaging. When contact mode is used, the setpoint is the force applied by the tip to the sample. Scan size determines the x-y dimensions of the scan. The points/line (pixel resolution) affects the digital resolution. Increasing the resolution will improve image quality but will require longer imaging times.

Scan speed controls the number of lines scanned per second. A faster scan speed decreases imaging time but may not allow the system sufficient time to trace surface topography. A scan speed of 2–5 lines/second is suitable for smooth surfaces, but rougher surfaces may require a lower scan speed [1].

End-of-Chapter Questions

1. AFM electronics are responsible for all of the following except:
   a. x-y piezo scanner control
   b. z-scanner control
   c. collection of signals for computer processing
   d. all of the above are controlled by AFM electronics

2. Digital signal processors (DSPs) are responsible for:
   Choose all that apply.
   a. process signals from the scanner head
   b. perform feedback control calculations

   c.   control x-y raster scan functions

   d.   laser alignment·

3. What is the role of the photodiode?

   a.   to align the laser

   b.   to convert laser light into an analog voltage signal

   c.   to produce a feedback loop signal

   d.   to position the tip on the sample surface

4. True or False. The photodiode can be used to monitor cantilever motion.

   a.   True

   b.   False

5. True or False. The photodiode can only monitor vertical motion of the cantilever.

   a.   True

   b.   False

6. What type of materials are found in AFM scanners?

   a.   electromechanical

   b.   bimetallic

   c.   piezoelectric

   d.   polymer

7. The expansion coefficient in scanner materials is:

   a.   10 nm per volt

   b.   5 nm per volt

   c.   1 nm per volt

   d.   0.1 nm per volt

8. The materials found in AFM scanners possess geometries that resemble a:

   a.   square

   b.   tube

   c.   rectangle

   d.   triangle

9. When the scanner responds slowly to an applied voltage, what scanner nonlinearity appears in the AFM image?

   a.   creep

   b.   hysteresis

10. Bending distortions appear in AFM images due to scanner _____.
    a. creep
    b. hysteresis

11. The _____ approach brings the tip and sample together from initial distances of 10s or 100s μm apart.
    a. fine
    b. coarse

12. The _____ approach brings the tip in contact with the sample surface.
    a. fine
    b. coarse

13. The feedback loop compares the photodiode signal to:
    a. the scan size
    b. the pixel resolution
    c. the setpoint
    d. the scan speed

14. The feedback loop maintains constant:
    a. laser intensity
    b. tip-sample forces
    c. photodiode output
    d. a and c

15. The error signal is used to:
    Choose all that apply.
    a. drive the z piezo
    b. control the x-y raster pattern
    c. generate sample topography
    d. control scan rate

16. If tip-sample forces increase, the feedback loop moves the tip _____ the surface.
    a. toward
    b. away from

17. If tip-sample forces decrease, the feedback loop moves the tip _____ the surface.
    a. toward
    b. away from

18. Which of the following parameters control how quickly feedback electronics respond to tip deflection?

    a. scan size

    b. pixel resolution

    c. setpoint

    d. gains

19. Which of the following controls probe movement over larger features on the sample surface?

    a. integral gains

    b. proportional gains

20. Which of the following controls probe movement over smaller, more frequent features on the sample surface?

    a. integral gains

    b. proportional gains

References

[1] P. Eaton and P. West, *Atomic Force Microscopy*, Oxford: Oxford University Press, 2010.

[2] F. Moreno-Herrero and J. Gomez-Herrero, "AFM: Basic concepts," in *Atomic Force Microscopy in Liquid—Biological Applications*, A. M. Baro and R. G. Reifenberger, Eds., Weinheim, Wiley-VCH Verlag GmbH & Co. KGaA., 2012, pp. 3–34.

[3] E. Thormann, T. Pettersson and P. M. Claesson, "How to measure forces with atomic force microscopy without significant influence from nonlinear optical lever sensitivity," *Rev. Sci. Instrum.*, vol. 80, pp. 093701-1–093701-11, 2009.

[4] K. D. Jandt, "Atomic force microscopy of biomaterials surfaces and interfaces," *Surf. Sci.*, vol. 491, pp. 303–332, 2001.

[5] R. R. De Oliveira, D. A. Albuquerque, T. G. Cruz, F. M. Yamaji and F. L. Leite, "Measurement of the nanoscale roughness by atomic force microscopy: Basic principles and applications," in *Atomic Force MIcroscopy—Imaging, Measuring, and Mainpulating Surfaces at the Atomic Scale*, V. Bellitto, Ed., Rijeka, InTech, 2012, pp. 147–174.

[6] K. M. Jeric, *An Experimental Evaluation of the Application of Smart Damping Materials for Reducing Structural Noise and Vibrations*, Blacksburg: Virginia Polytechnic Institute and State University, 1999.

[7] Y. Li, L. Zhang, G. Shan, Z. Song, R. Yang, H. Li and J. Qian, "A homemade atomic force microscope based on a quartz tuning fork for undergraduate instruction," *Am. J. Phys.*, vol. 84, pp. 478–485, 2016.

[8] G. J. Vancso and H. Schonherr, "Atomic force microscopy in practice," in *Scanning Force Microscopy of Polymers*, Berlin, Springer-Verlag, 2010, pp. 25-27.

4

AFM Cantilevers and Probes

4.0 Key Objectives

- Become familiar with the components of AFM probes.
- Learn common probe dimensions.
- Understand the importance of the following cantilever mechanical properties:
  - spring constant
  - resonant frequency
- Learn fabrication techniques used to make AFM probes.

4.1 Probe Characteristics

AFM probes consist of a tip, millimeter-scale handle, and a spring-like cantilever (Figure 4.1) [1]. In order to make handling simple, tip manufacturers attach cantilever beams to substrates, also known as chips. Substrates are usually 3.5 mm × 1.6 mm × 0.5 mm thick to accommodate the mounting of probes from various manufacturers in different AFM probe holders [2].

4.2 Tip Geometry

The standard, pyramid-shaped AFM tip has a radius of curvature of 5–10 nm for Si and 20–60 nm for Si_3N_4, because tip radii should be as small as possible [3]. Typical half cone angles of pyramidal AFM tips range from 10° to 35° (Figure 4.2) [4–6]. The geometry of the probe tip used to scan the surface AFM governs image resolution. For this reason, accurate knowledge of the AFM tip shape is critical for accurate dimensional measurements [1]. The geometry of

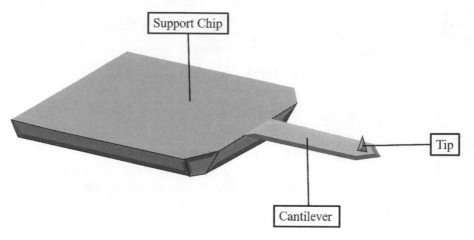

FIGURE 4.1
The key components of an AFM probe.

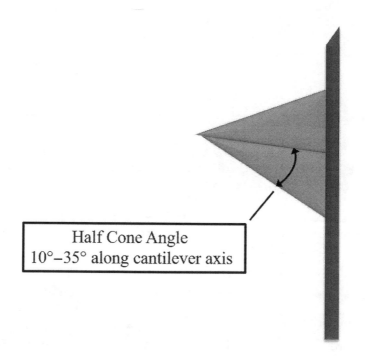

FIGURE 4.2
Half cone angle.

the probe has a dramatic effect on horizontal resolution. Sharper tips produce higher horizontal resolution [2]. A tip that is smaller than the surface features will more accurately reproduce features on the sample surface (Figure 4.3a) than a blunt tip (Figure 4.3b). Images produced with a tip whose geometrical features are larger than the surface features will produce image features that are significantly wider than the true surface geometry [6].

Particles appear wider with blunt probes (Figure 4.4a). Pits appear less wide and less deep (Figure 4.4b). If the probe cannot reach the bottom of a surface pit or track the sides of a particle, the image will not indicate the correct geometry of the sample. Blunt probes will lead to images with features larger than expected [2].

Several techniques exist for the improvement of tip geometry. Si_3N_4 tips can be sharpened to improve their aspect ratio and reduce the tip radii to as low as 5 nm. Si tips can be machined with focused ion beam (FIB) instruments to improve the aspect ratio. Si tips are now available with radii less than 5 nm [3].

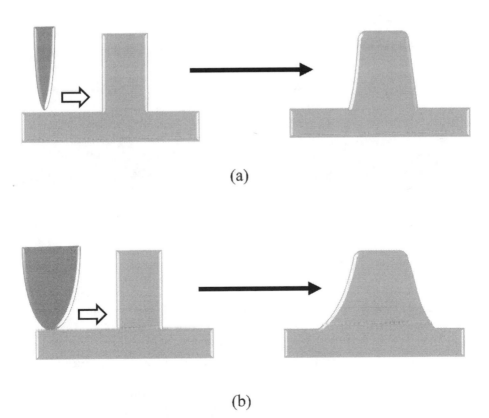

(a)

(b)

FIGURE 4.3
Cross-sectional profiles of a step imaged with a sharp tip (a) and a blunt tip (b).

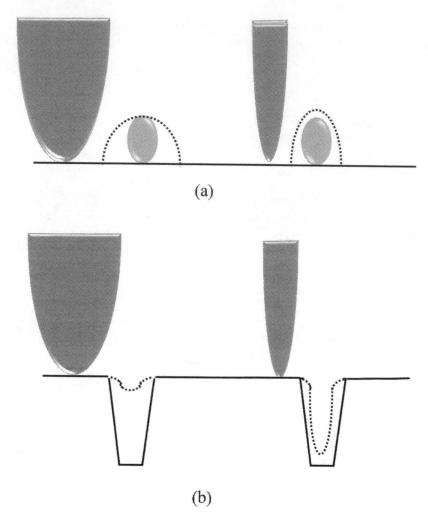

FIGURE 4.4
Blunt and sharp AFM tips imaging particles (a) and pits (b).

4.3 Cantilever Characteristics

The cantilever is a key element of the AFM and its mechanical properties are largely responsible for its performance [7]. Cantilevers are mechanical devices specially designed to measure tiny forces [4]. Tip deflections, angstroms in size, which corresponds to forces of approximately 10 pN, can be reliably measured using standard cantilevers [3]. In order to register a measurable deflection with small forces, the cantilever must flex with a relatively low force on the nN scale [4].

Si_3N_4 cantilevers are less expensive than those made of other materials. They are very rugged and well suited to imaging in almost all environments [4]. The cantilever stylus used in the AFM should meet the following criteria: (1) low normal spring constant, (2) high resonance frequency, (3) high cantilever quality factor Q, (4) high lateral spring constant, (5) short cantilever length, (6) incorporation of reflective coatings for deflection sensing, and (7) a sharp protruding tip [4]. Selection of the appropriate tip and cantilever depends on the application. The use of soft cantilevers in contact mode prevents sample damage. Stiff cantilevers are required for imaging in tapping mode to overcome capillary forces [3]. Cantilever length ranges from 100 to 200 μm, widths fall between 10 to 40 μm, and thicknesses range from 0.3 to 2 μm [8]. The most common cantilever length is 115 μm long [8]

4.4 Mechanical Properties of Cantilevers

The spring constant and the resonance frequency are the key mechanical properties of cantilevers. Both can be determined using the material properties and dimensions of the cantilever (Figure 4.5) [8].

4.4.1 Spring Constants

The nature of the AFM experiment determines the type of cantilever required. When a different stiffness is required, switching the cantilevers and repeating the laser alignment procedure is necessary. Two basic AFM operating modes are contact mode and tapping mode. The cantilevers used for contact mode have spring constants that are typically much less than 1 N/m and are made from Si_3N_4 [2]. A cantilever with an extremely low spring constant is required for high vertical and lateral resolutions at small forces (0.1 nN or lower), but a high resonant frequency is desirable (about 10–100 kHz) at the same time in order to minimize the sensitivity to external, mechanical vibrations. This requires a spring with an extremely low vertical spring constant (typically 0.05–1 N/m) [4]. The spring constant of the cantilever depends on its shape, its dimensions, and composition [4]. Equation 4.1 mathematically describes the spring constant:

$$k = \frac{Et^3w}{4l^3} \tag{4.1}$$

where E is the Young's modulus of the material (i.e., for Si_3N_4 $E = 1.5 \times 10^{11}$ N m^{-2}) and t, w, and l are the thickness, width, and length of the cantilever, respectively [9]. Additionally, a cantilever has lateral and torsional spring constants [2]. An explanation of torsional spring constant importance is in Chapter 6.

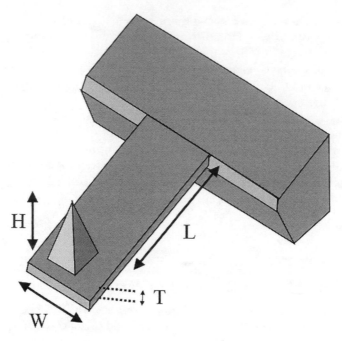

FIGURE 4.5
Cantilever dimensions.

4.4.2 Resonance Frequency

Resonance frequencies range from a few kilohertz to hundreds of kilohertz. These frequencies provide high-speed responses for tapping mode operation [8]. The fundamental frequency (ω_0) of any spring is given by Equation 4.2:

$$\omega_0 = \frac{1}{2\pi}\sqrt{\frac{k}{m_{eff}}} \tag{4.2}$$

where k is the spring constant (stiffness) in the normal direction and m_{eff} is the effective mass [4]. In the case of AFM cantilevers, given the modulus of elasticity (E), cantilever width (b), height (thickness) of the cantilever (h), length of the cantilever beam (L), and the mass of the cantilever, it is possible to determine the resonant frequency as shown in Equation 4.3 [8]:

$$\omega_0 = \sqrt{\frac{Ebh^3}{4L^3 m_{eff}}} \tag{4.3}$$

4.5 Probe Fabrication

Fabrication of Si and Si_3N_4 AFM cantilevers involves photolithographic techniques [4]. Processes for tip fabrication starts with formation of a tip mold using anisotropic silicon etching with silicon dioxide as an etch-mask. This produces a pyramid-shaped mold used with successive wet etch, metal deposition, and photolithographic steps as shown in Figure 4.6a-g [10].

Improvements in AFM performance is possible with sharper tips. There are several techniques available for sharpening AFM tips in order to improve horizontal resolution [2]. In addition to using photolithographic methods for tip fabrication, direct writing methods using focused ion beam (FIB) allows precise nanopatterning and precise material processing with high accuracy.

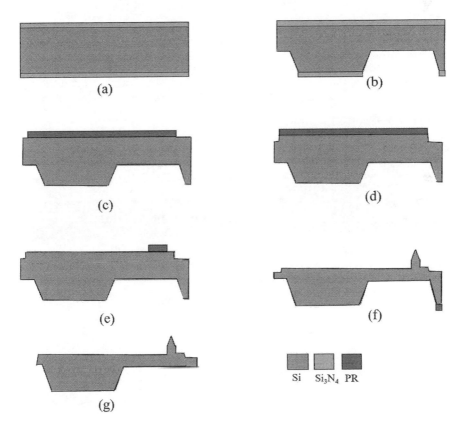

FIGURE 4.6
Fabrication of AFM probes starting with deposition of Si_3N_4 (a), followed by etching with potassium hydroxide (b). Photolithography is used to pattern the cantilever structure (c) followed by another etch step to set cantilever thickness (d). Tip mask patterning (e), tip etching (f), and probe release (g) end the process.

The characteristic feature of FIB is the high spatial resolution achieved due to the use of a gallium ion beam 7 nm in diameter [11]. Tip sharpening is also possible with chemical etching, ion milling, or adding a carbon nanotube. Each of these techniques can create a sharper probe, but also add to the price of fabricating the probe [2].

End-of-Chapter Questions

1. The main components of an AFM probe include all of the following except:
   a. substrates
   b. lasers
   c. cantilevers
   d. tips

2. The dimensions of the AFM support chip (substrate) are generally:
   a. 4.5 mm × 2.6 mm × 1.5 mm
   b. 3.5 mm × 1.6 mm × 0.5 mm
   c. 2.5 mm × 0.6 mm × 2.5 mm
   d. 1.5 mm × 2.6 mm × 1.5 mm

3. Radius of curvature for AFM tips is:
   a. 80–100 nm
   b. 40–60 nm
   c. 10–20 nm
   d. 5–10 nm

4. Half cone angles of pyramidal AFM tips ranges from:
   a. 10–35°
   b. 5–10°
   c. 40–50°
   d. 60–80°

5. When a blunt AFM tip is used to scan sample surfaces, particles appear _____.
   a. larger
   b. smaller

6. True or False. When a blunt AFM tip is used to scan sample surfaces, pits appear less wide and less deep.

   a. True

   b. False

7. Blunt probes will produce images with features _____ than expected.

   a. larger

   b. smaller

8. Contact mode imaging requires _____ cantilevers and tapping mode imaging requires _____.

   a. soft, stiff

   b. stiff, soft

9. Cantilever length ranges from _____ μm to _____ μm.

   a. 10, 20

   b. 100, 200

   c. 200, 300

   d. 400, 500

10. Cantilever spring constants depend on all of the following except:

   a. reflectivity

   b. shape

   c. dimensions

   d. composition

References

[1] S. Zheng, C. Zhu, R. Kumar and B. Cui, "Batch fabrication of AFM probes with direct positioning capability," *J. Vac. Sci. Technol. B*, vol. 35, pp. 06GC02-1–06GC02-4, 2017.

[2] P. Eaton and P. West, *Atomic Force Microscopy*, Oxford: Oxford University Press, 2010.

[3] J. H. Hafner, C. L. Cheung, A. T. Woolley and C. M. Lieber, "Structural and functional imaging with carbon nanotube AFM probes," *Prog. Biophys. Mol. Biol.*, vol. 77, pp. 73–110, 2001.

[4] B. Bhushan and O. Marti, "Scanning Probe Microscopy—Principle of Operation, Instrumentation, and Probes," in *Nanotribology and Nanomechanics*, B. Bhushan, Ed., Berlin, Springer-Verlag, 2005, pp. 41–115.

[5] L. Chen, C. L. Cheung, P. D. Ashby and C. M. Lieber, "Single-walled carbon nanotube AFM probes: Optimal imaging resolution of nanoclusters and biomolecules in ambinet and fluent environments," *Nano Lett.*, vol. 4, pp. 1725–1731, 2004.

[6] E. E. Flater, G. E. Zacharakis-Jutz, B. G. Dumba, I. A. White and C. A. Clifford, "Toward easy and reliable AFM tip shape determination using blind tip reconstruction," *Ultramicroscopy*, vol. 146, pp. 130–143, 2014.

[7] H. J. Butt, B. Cappella and M. Kappl, "Force measurements with the atomic force microscope: Technique, interpretation and applications," *Surf. Sci. Rep.*, vol. 59, pp. 1–152, 2005.

[8] R. R. De Oliveira, D. A. Albuquerque, T. G. Cruz, F. M. Yamaji and F. L. Leite, "Measurement of the nanoscale roughness by atomic force microscopy: Basic principles and applications," in *Atomic Force MIcroscopy—Imaging, Measuring, and Mainpulating Surfaces at the Atomic Scale*, V. Bellitto, Ed., Rijeka, InTech, 2012, pp. 147–174.

[9] F. Moreno-Herrero and J. Gomez-Herrero, "AFM: Basic concepts," in *Atomic Force Microscopy in Liquid—Biological Applications*, A. M. Baro and R. G. Reifenberger, Eds., Wiley-VCH Verlag GmbH & Co. KGaA., 2012.

[10] J. P. Rasmussen, P. T. Tang, C. Sander, O. Hansen and P. Moller, "Fabrication of an all-metal atomic force microscope probe," *Proc. Transduc.*, vol. 1, pp. 463–466, 1997.

[11] O. A. Ageev, A. S. Kolomiytsev, A. V. Bykov, V. A. Smirnov and I. N. Kots, "Fabrication of advanced probes for atomic force microscopy using focused ion beam," *Microelectron. Reliab.*, vol. 55, pp. 2131–2134, 2015.

5

Contact Mode AFM

5.0 Key Objectives

- Become familiar with contact mode AFM applications.
- Understand the forces exerted between the tip and sample during contact mode operation.
- Understand feedback loop during contact mode operation.
- Become familiar with surface roughness analysis.

5.1 Contact Mode Characteristics

Contact mode atomic force microscopy is the oldest and simplest AFM imaging mode in which the tip is in direct contact with the surface [1]. Contact mode is also known as static or repulsive mode [2]. In contact mode, the tip remains in contact with the sample surface while scanning in a raster pattern [3]. It is the fastest of all topographic modes. Monitoring cantilever deflection allows production of topographical images [4].

5.1.1 Contact Mode Applications

Contact mode imaging is best suited for relatively flat and hard samples since the tip applies large lateral forces [3]. The most frequent use of contact mode involves imaging hard inorganic surfaces, metals, and for high molecular weight polymers. Additional samples imaged with contact mode AFM include fixed cells, proteins, cell surfaces, or low modulus biomaterials [5]. Attempts to image isolated biological structures usually result in poor resolution, sample movement, or damage [3]. The height images generated in contact mode provide information about quantitative height topography, surface roughness, and thickness of biological layers, while the deflection image reveals fine surface details [6].

5.2 Probe Behavior in Contact Mode

When the tip is far from the sample surface, the cantilever has zero deflection. As the instrument continues to push the cantilever toward the surface, the interaction moves into the "repulsive" regime. The tip is now applying a force to the sample, and the sample applies an opposite force to the tip [4]. The atoms at the end of the tip experience a repulsive force due to electronic orbital overlap with the atoms in the surface of the sample [2]. The force acting on the tip causes the cantilever to deflect (Figure 5.1). The optical detection scheme used to measure cantilever deflection involves a laser and a photodiode, as described in Section 3.2. Deflection can be measured to within 0.02 nm, so a force as low as 0.2 nN can be detected [2].

5.3 Feedback Loop Operation in Contact Mode

A feedback mechanism measures and maintains constant cantilever deflection while the tip scans the sample surface [7]. The z position of the scanner adjusts during scanning to maintain constant cantilever deflection (i.e., constant force) at a user-defined setpoint value [3]. The feedback loop, described in Section 3.3, controls z piezo adjustment and provides data used to generate a topographic image of the sample surface [3].

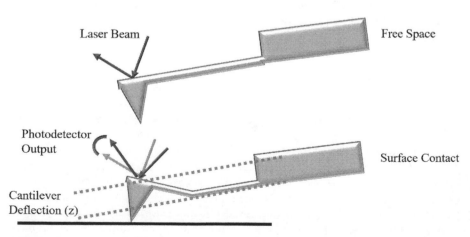

FIGURE 5.1
Cantilever deflection above the sample surface and in contact with the sample surface.

5.3.1 Setpoint

The basis of contact-mode AFM is that the microscope feedback system acts to keep the cantilever deflection at a constant value determined by the AFM operator. This value is the setpoint [4]. The setpoint is the force between the tip to the sample. The feedback circuits acts to keep cantilever deflection at the user-defined setpoint value [4].

5.3.2 Deflection

The deflection signal in contact-mode AFM is the signal that indicates the extent of the cantilever deflection before correction by the feedback circuit via height adjustment by the z piezo [4]. A direct relationship exists between cantilever deflection and the tip-sample interaction force. A constant value of cantilever deflection selected by the user (setpoint). As deflection is maintained by the feedback loop, the force between the tip and sample is kept constant [4]. Because the feedback system of the AFM cannot have instantaneous response, the vertical deflection will actually vary somewhat during imaging. The amount it varies will depend on the topography of the sample, flexibility of the cantilever, scanning speed, and how well the feedback circuit has been optimized [4].

5.3.3 Feedback Loop Signals

The feedback signal in contact mode is the photodiode response that corresponds to vertical deflection of the cantilever [8]. The feedback loop compares the deflection signal to the setpoint and changes the voltage applied to the z piezo of the scanner, to maintain constant cantilever deflection as shown in Figure 5.2 [9]. As the tip scans the surface, the z piezo adjusts to maintain the signal from the photodiode equal to the setpoint [8]. Height data used to create topographical images comes from the voltage sent to the z piezo [5].

5.4 Surface Roughness

5.4.1 Average Roughness

Average roughness (R_a), the most common surface roughness measurement, is the mean height calculated over a given length. R_a describes the roughness of machined surfaces and provides AFM users the ability to detect general variations in profile height characteristics [6, 10]. Mathematically, R_a is the arithmetic mean of the absolute values of the height of the surface profile, as shown in Equation 5.1:

$$R_a = \frac{|M_1 + M_2 + M_3 \ldots M_N|}{N} \tag{5.1}$$

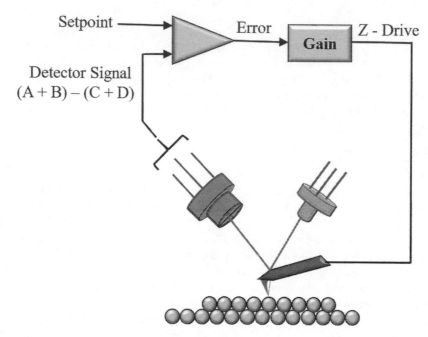

FIGURE 5.2
The feedback loop sending signals to the z piezo to maintain a constant setpoint.

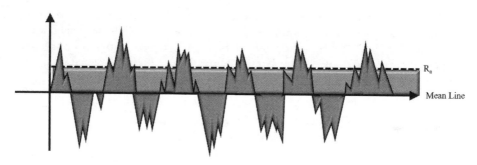

FIGURE 5.3
Average roughness (R_a) based on the mean line.

N represents the total data points involved in the measurement, and M is the vertical deviations measured from the average height (mean line) of the surface as shown in Figure 5.3 [6]. Height and spacing parameters are the physical parameters that govern surface roughness (Figure 5.4). The roughness average (R_a) is the most widely used because it is a simple parameter to obtain.

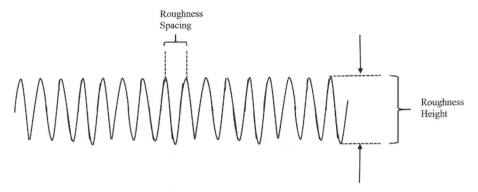

FIGURE 5.4
Height and spacing parameters used in average roughness determinations.

The average roughness makes no distinction between peaks and valleys. It becomes a disadvantage to characterize an average surface roughness if this data is relevant. Average roughness can be similar for surfaces with completely different roughness profiles because it depends only on the average profile of heights. For this reason, there is a need for parameters that are more sophisticated for a complete characterization of surfaces, and to distinguish between peaks and valleys [11].

5.4.2 RMS Roughness

The root mean square (RMS) roughness (R_q) is more sensitive to peaks and valleys than the average roughness due to the squaring of the measurements in the calculation [11]. R_q roughness is the square root of the distribution of surface height and is more sensitive than the average roughness for large deviations from the mean line/plane [10]. This parameter is also more sensitive to large deviation from the mean line as opposed to R_a (Figure 5.5). The mathematical definition of R_q appears in Equation 5.2:

$$R_q = \frac{\sqrt{M_1^2 + M_2^2 + M_3^2 \ldots M_N^2}}{N} \qquad (5.2)$$

where N refers to the number data points while M is the vertical deviation with respect to the mean line [6].

5.4.3 Additional Roughness Parameters

Roughness skewness (R_{sk}) measures the symmetry of the variations of a profile/surface about the mean line/plane and is more sensitive to occasional

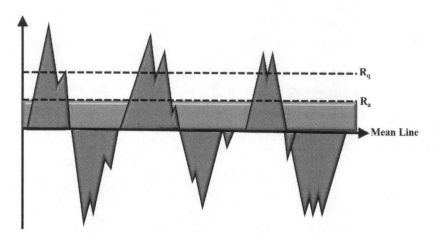

FIGURE 5.5
RMS roughness (Rq) compared to average roughness (Ra) and the mean line.

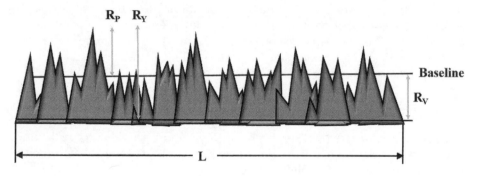

FIGURE 5.6
Illustration depicting the maximum height profile (R_Y), maximum profile valley depth (R_V), and maximum profile peak height (R_P).

deep valleys or high peaks. Roughness kurtosis (R_{ku}) is a measure of the distribution of the spikes above and below the mean line/plane. Usually, R_{sk} distinguishes two profiles of the same R_a or R_q values but of different shapes [10]. The maximum profile peak height (R_p) is the measure of the highest peak around the surface profile from the baseline. The maximum profile valley depth (R_v) is the measure of the deepest valley across the surface profile analyzed from the baseline. The maximum height of the profile (R_y) is the vertical distance between the deepest valley and highest peak [11]. All three appear in Figure 5.6.

5.5 Laboratory Exercise: Surface Roughness Analysis of Metallized Polymer Patterns

5.5.1 Laboratory Objectives

The laboratory objectives covered in this laboratory exercises include:

- preparation of polymer and metal nanogrids
- execution of AFM imaging to collect contact mode images
- determination of roughness parameters

5.5.2 Materials and Procedures

The recommended samples for use in this experiment are silver nanogrids produced from the metallization of polyvinylpyrrolidone (PVP) templates using dilute concentrations of silver nitrate and sodium citrate as a reducing agent. This silver nanogrid fabrication is reported in the literature [12]. Use double-sided tape to mount samples to 2 cm imaging discs. The AFM software used in this lab should have surface roughness as part of the analysis options. Substrate preparation involves cutting 2 cm × 2 cm glass square from a clean microscope slides. Polymer templates are microcontact printed on the glass squares following the procedures in the literature [13]; these samples serve as the control. The polymer templates are metallized following procedures stated in the literature [12]. Use double-sided tape to mount the glass squares containing the polymer grids and silver nanogrids on 2 cm imaging discs. Prepare the AFM system for contact mode imaging according to the AFM manual.

5.5.3 Sample Data

Figure 5.7a shows a contact mode AFM image of a polymer grid, and Figure 5.7b presents a contact mode image of a silver nanogrid. These images were prepared using free image processing software [14].

Table 5.1 show roughness data for polymer grids and silver nanogrids.

Post-Lab Questions

1. Surface roughness was more prominent with the _____.
   a. polymer grid
   b. silver nanogrid

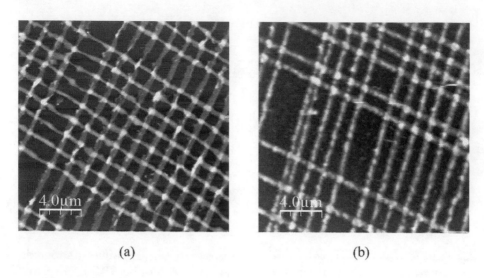

(a)   (b)

FIGURE 5.7
Contact mode image of polymer grid (a) and silver nanogrid (b).

TABLE 5.1

Surface Roughness Values for Polymer and Silver Nanogrids

Property	Polymer Frid	Silver Nanogrid
Average Roughness (R_a)	7.7785 nm	21.256 nm
RMS Roughness (R_q)	10.158 nm	24.908 nm
Peak-Valley Height (R_y)	42.71 nm	78.16 nm
Peak Height (R_p)	9.9078 nm	44.766 nm

2. When comparing the polymer grid and the silver nanogrid, an increase in surface roughness was observed:
   a. with R_a only
   b. with R_q only
   c. with R_a and R_q
   d. surface roughness did not change

3. Metallization of the polymer grid resulted in an increase in:
   a. R_y only
   b. R_p only
   c. R_y and R_p
   d. none of the above

4. Consider the polymer grid and silver nanogrid samples. Which of the following was a more sensitive measure of surface roughness?

    a. R_a

    b. R_q

5. Which of the following confirms the deposition of a metal coating on the polymer template?
    Choose all that apply.

    a. R_a

    b. R_q

    c. R_y

    d. R_p

End-of-Chapter Questions

1. Contact mode AFM imaging is also referred to as:
    Choose all that apply.

    a. dynamic mode

    b. static mode

    c. repulsive mode

    d. phase

2. During contact mode imaging, cantilever _____ is monitored to produce topographical images.

    a. twisting

    b. deflection

3. Contact mode imaging is the preferred method for which sample type?
    Choose all that apply.

    a. biological

    b. soft polymers

    c. flat

    d. hard

4. True or False. Hard molecular weight polymers are frequently imaged using contact mode.

    a. True

    b. False

5. While imaging in contact mode, _____ images provide topography data, surface roughness, and film thickness information. _____ images reveal fine surface detail.

   a. deflection, height

   b. height, deflection

6. When the electronic orbital overlap occurs between tip atoms and sample atoms _____ forces are observed.

   a. attractive

   b. repulsive

7. The feedback loop adjusts the ____ position of the scanner to maintain constant cantilever deflection during contact mode imaging.

   a. x

   b. y

   c. z

8. The feedback loop signal is used to generate _____ images.

   a. topographical

   b. deflection

9. Which of the following parameters determines the tip-sample force maintained during contact mode imaging?

   a. scan speed

   b. setpoint

   c. gains

   d. pixel resolution

10. The feedback loop compares the setpoint to the _____ to determine how much to raise or lower the scanner to maintain constant cantilever deflection.

    a. laser intensity

    b. vibrational amplitude

    c. setpoint

    d. scan size

11. Height data used to create topographical images comes from the voltage sent to the ____ piezo.

    a. x

    b. y

    c. z

12. Which of the following is the most common surface roughness measurement?

   a. R_a
   b. R_q
   c. R_y
   d. R_p

13. Which of the following is the arithmetic mean of the absolute values of the heights relative to a baseline?

   a. R_a
   b. R_q
   c. R_y
   d. R_p

14. Which of the following measurements makes no distinction between peaks and valleys?

   a. R_a
   b. R_q

15. Which of the following is more sensitive to peaks and valleys?

   a. R_a
   b. R_q

16. Which of the following is used to distinguish two profiles with the same R_a or R_q values?

   a. R_y
   b. R_p
   c. R_{sk}
   d. R_{ku}

References

[1] G. Zavala, "Atomic force microscopy, a tool for characterization, synthesis and chemical processes," *Colloid Polym. Sci.*, vol. 286, pp. 85–95, 2008.

[2] B. Bhushan and O. Marti, "Scanning probe microscopy—Principle of operation, instrumentation, and probes," in *Nanotribology and Nanomechanics*, B. Bhushan, Ed., Berlin, Springer-Verlag, 2005, pp. 41–115.

[3] J. H. Hafner, C. L. Cheung, A. T. Woolley and C. M. Lieber, "Structural and functional imaging with carbon nanotube AFM probes," *Prog. Biophys. Mol. Biol.*, vol. 77, pp. 73–110, 2001.

[4] P. Eaton and P. West, *Atomic Force Microscopy*, Oxford: Oxford University Press, 2010.

[5] N. H. Pedro, E. R. López, B. Pineda and J. Sotelo, "Atomic force microscopy in detection of viruses," in *Atomic Force Microscopy Investigations into Biology—From Cell to Protein*, C. L. Frewin, Ed., Rijeka, InTech, 2012, pp. 235–252.

[6] M. Marrese, V. Guarino and L. Ambrosio, "Atomic force microscopy: A powerful tool to address scaffold design in tissue engineering," *J Funct. Biomater.*, vol. 8, pp. 1–20, 2017.

[7] F. M. Herrero, J. Colchero, J. G. Herrero and A. M. Baro, "Atomic force microscopy contact, tapping, and jumping modes for imaging biological samples in imaging biological samples in liquids," *Phys. Rev. E*, vol. 69, pp. 031915-1–031915-9, 2004.

[8] F. Moreno-Herrero and J. Gomez-Herrero, "AFM: Basic Concepts," in *Atomic Force Microscopy in Liquid—Biological Applications*, A. M. Baro and R. G. Reifenberger, Eds., Wiley-VCH Verlag GmbH & Co. KGaA., 2012.

[9] G. Zavala, "Atomic force microscopy, a tool for characterization, synthesis, and chemical processes," *Colloid Polym. Sci.*, vol. 286, pp. 85–95, 2008.

[10] B. J. Kumar and T. S. Rao, "AFM studies on surface morphology, topography, and texture of nanostructured zinc aluminum oxide thin films," *Dig. J. Nanomater. Biostruct.*, vol. 7, pp. 1881–1889, 2012.

[11] R. R. De Oliveira, D. A. Albuquerque, T. G. Cruz, F. M. Yamaji and F. L. Leite, "Measurement of the nanoscale roughness by atomic force microscopy: Basic principles and applications," in *Atomic Force MIcroscopy—Imaging, Measuring, and Mainpulating Surfaces at the Atomic Scale*, V. Bellitto, Ed., Rijeka, InTech, 2012, pp. 147–174.

[12] W. C. Sanders, R. Valcarce, P. Iles, J. S. Smith, G. Glass, J. Gomez, G. Johnson, D. Johnston, M. Morham, E. Befus, A. Oz and M. Tomaraei, "Printing silver nanogrids on glass," *J. Chem. Educ.*, vol. 94, pp. 758–763, 2017.

[13] W. C. Sanders, "Fabrication of polyvinylpyrrolidone micro-/nanostructures utilizing microcontact printing," *J. Chem. Educ.*, vol. 92, pp. 1908–1912, 2015.

[14] I. Horcas, R. Fernandez, J. M. Gomez-Rodriguez, J. M. Colchero, J. Gomez-Herrero and A. M. Baro, "WSXM: A software for scanning probe microscopy and tool for nanotechnology," *Rev. Sci. Instrum.*, vol. 78, pp. 013705-1–013705-8, 2007.

6

Lateral Force Microscopy

6.0 Key Objectives

- Understand the primary uses of lateral force microscopy.
- Understand the basic operation of lateral force microscopy.
- Learn how to interpret friction loop data.

6.1 Introduction

When the probing tip slides along the surface, frictional forces arise. Microscopes with the capability of measuring both topography and friction channels simultaneously are lateral force microscopes (LFMs), also known as frictional forces microscopes (FFMs) [1]. Three main uses of lateral force microscopy include fundamental study of tribology of macroscopic or atomic scale systems (1), determination of the friction properties of uniform materials (2), and characterization of heterogeneous materials based on their frictional properties (3) [2]. LFMs can distinguish frictional properties of material at the nanoscale. A common LFM use involves detection of features deposited by dip-pen nanolithography because LFM can discriminate between different chemical functionalities [2]. LFMs allow the study and understanding of nanotribological phenomena of the surfaces at the nanometer scale, such friction, wear, and lubrication [3]. LFM is particularly useful when studying heterogeneous materials on the nanoscale because LFM reveals tribological contrasts caused by material differences [4, 5]. These contrasts arise from differences in chemical interactions between different molecular regions [5]. It is interesting to note that modification of the LFM tip with functional groups allows hydrophobic or hydrophilic areas on the sample surfaces to become visible with nanometer resolution. Similar chemical groups at the tip and the sample surface interact more than dissimilar groups. A stronger interaction leads to a higher friction between the tip and the sample [5]. What makes this technique unique is its ability to distinguish frictional properties of material at the nanoscale. This means that differentiation of materials is possible based on their frictional properties with the potentially atomic-scale. This makes it possible to perform compositional mapping with LFM [4].

6.2 Lateral Force Microscope Probe Behavior

Twisting of the cantilever due to changes in tip-sample interactions changes the position of the laser beam on the photodiode [6]. This allows LFMs to monitor vertical bending and torsion of the cantilever simultaneously while operating in contact mode [7]. The magnitude of cantilever twisting is proportional to the friction encountered as the tip scans over the sample. The cantilever will twist by a certain amount if there is measurable friction present (Figure 6.1) [2]. The twisting motion reflects the frictional properties of the surface, obtained by monitoring the photodiode [8]. Differences in the friction coefficient will produce different torsions of the cantilever [9].

In LFM the cantilever tip remains in permanent contact with the sample surface while being rastered over the surface at constant velocity [2]. Two more considerations should be mentioned: during scanning, the torsion

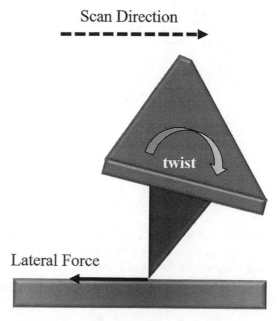

FIGURE 6.1
Twisting of the cantilever due to lateral forces in LFM.

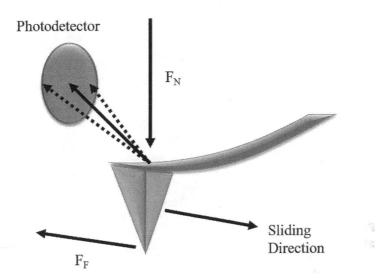

FIGURE 6.2
The shear and normal forces involved with lateral force microscopy.

of the cantilever is induced by lateral forces acting in the x-direction (F_x), whereas for corrugated samples, the bending of the cantilever results from forces acting perpendicular to the sample surface in the z-direction (F_z) (Figure 6.2). Therefore, on the atomic scale, the lateral forces in the same scan direction and perpendicular to the scan direction are measured [7].

6.3 Lateral Force Microscopy Data

The frictional force (F_f) depends on the normal force (F_n) applied. For many materials the relationship will be linear, meaning that a plot of normal versus lateral forces allows the calculation of the useful parameter μ, the friction coefficient of the material [4]. The relationship between F_f, F_n, and μ is shown in Equation 6.1 [2]:

$$F_f = \mu \bullet F_n \qquad (6.1)$$

6.3.1 Photodiode Response in Lateral Force Microscopy

Monitoring the lateral forces that torque the tip, causing the cantilever to twist, takes place by observing the horizontal movement of the laser spot (Figure 6.3) [2].

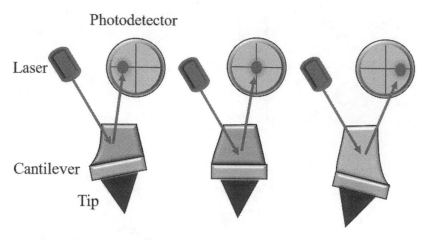

FIGURE 6.3
Movement of laser beam across the photodiode.

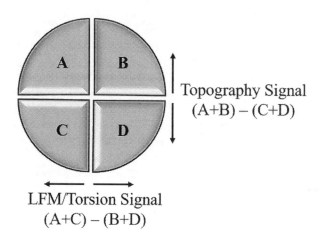

FIGURE 6.4
Topographical and frictional photodiode signals.

AFM systems have a quadrant photodiode, a detector divided into four quadrants. Current on each quadrant can be isolated or combined. In this type of systems, signal $(A + C) - (B + D)$ is a measurement of how much the cantilever is twisted, as shown in Figure 6.4 [9]. Comparison of the left- and right-hand sides of the photodiode allows measurement of lateral deflection signals. The lateral deflection signal will always be different in the forward and reverse scans (Figure 6.5). Even on perfectly flat, homogeneous samples, two images will differ in intensity and sign. The actual friction measured is the difference between the forward and reverse scans [2].

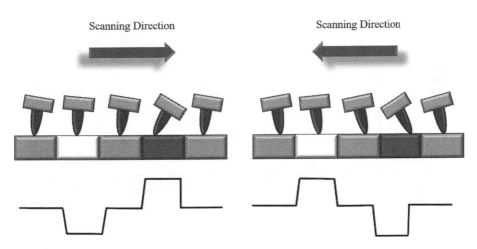

FIGURE 6.5
Forward and reverse scans in lateral force microscopy.

The signal difference between the upper and the lower halves of the photodiode is the *normal deflection signal*, a measure of the tip-sample force normal to the sample. Similarly, the signal difference between the left and the right halves of the diode, *the lateral deflection signal*, refers to the measure of the tip-sample interaction force along the sample surface. This lateral interaction force acts on the cantilever tip which results in its twisting, which in turn manifests as the lateral deflection signal [10]. The signal difference between the left and right sides of the photodiode provides lateral force data [6].

6.3.2 Friction Loop

The term "friction loop" refers to the combination of the LFM data from the forwards and reverse scans (Figure 6.6). Calculating the difference between forward and backwards scans provides the actual friction of the material under study. This calculated value may be presented as volts or converted to force [4]. The lateral deflection signal is different in the two directions, as the cantilever will twist by a certain amount assuming there is some measurable lateral component to the tip-sample force. Changes in friction due to material contrast will give greater or smaller difference between the forward and reverse scans. Larger friction will give a greater difference between the forward and reverse scans, while lower friction will give a smaller difference [2]. Subtracting the lateral force reverse trace from the forward trace provides qualitative friction data [11].

Figure 6.7 illustrates a typical friction loop from a LFM scan. At the turning points of the forward and reverse scans, a linear relationship (S) is observed between the photodiode signal and tip movement in the x direction [11].

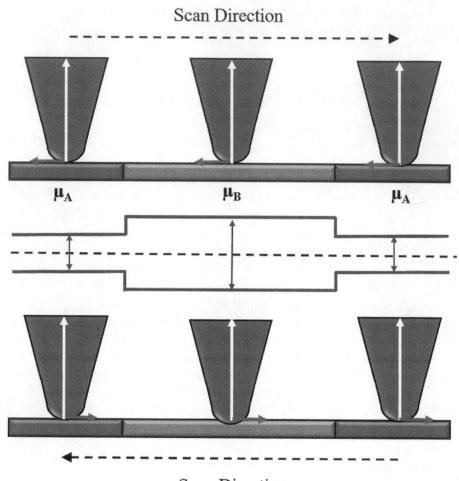

FIGURE 6.6
Friction loop examples.

For quantification, the torsional spring constant (k_t) of the cantilever, which is dependent on its dimensions, must be determined [12]. Equation 6.2 gives the torsional spring constant (k_t) for a rectangular cantilever:

$$k_t = w \times \frac{G}{3} \times \frac{t^3}{l} \times \frac{1}{(H + \frac{t}{2})^2} \tag{6.2}$$

where G is the modulus of rigidity (shear modulus), t is cantilever thickness, w is cantilever width, l is cantilever length, and H is tip height [4]. Long

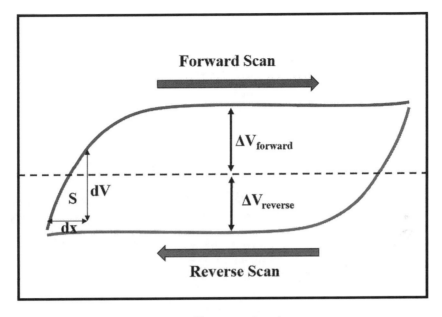

x distance (μm)

FIGURE 6.7
Friction loop illustrating photodiode sensitivity, S.

cantilevers with small thicknesses provide maximum torsion. [4]. Finally, the frictional force (F_f) can be determined using the friction loop signal (ΔV), the photodiode sensitivity (S), and the lateral spring constant (k_t) in accordance with Equation 6.3 [13]:

$$F_f = \frac{\Delta V \times k_t}{S} \qquad (6.3)$$

6.4 Laboratory Exercise: Analysis of Nanoshaved Patterns Etched in Polymer Films

6.4.1 Laboratory Objectives

The objectives of this experiment include:

- nanofabrication with the AFM tip
- acquisition of a lateral force microscope image
- acquisition of forward and reverse friction loops

TABLE 6.1

Recommended Nanoshaving Parameters

Parameters	Value
Nanoshaving Setpoint	30 nN
Nanoshaving Scan Window	5 μm
Nanoshaving Passes	3
Imaging Setpoint	10 nN

6.4.2 Materials and Procedures

Use a diamond scribe to cut 2 cm × 2 cm glass squares from a clean microscope slide. For this experiment, deposition of the polymer film involves the use of low-voltage laptop fans to spin coat water-based polyvinylpyrrolidone (PVP) films on glass substrates. Double-sided tape sufficiently anchors the glass substrates to the low-voltage laptop fans during spin coating. Prepare a 5 mg/mL aqueous PVP solution and use a plastic pipette to transfer the solution to the glass substrate attached to the laptop fan. Operate the laptop fan for approximately 30 seconds, and repeat the process two more times, resulting in three depositions of PVP on the glass substrate. Prepare the AFM for contact mode imaging according to the AFM manual. Use the AFM operating in contact mode to perform nanoshaving in order to etch a square in the PVP film [14]. Table 6.1 shows recommended nanoshaving parameters. After nanoshaving is complete, increase the scan size to 30 μm and scan the pattern using the imaging setpoint indicated in Table 6.1. Use two imaging windows to collect topographical and frictional data simultaneously. Use image processing software [15] to process images and generate friction loops.

6.4.3 Sample Data and Results

In the lateral force image (Figure 6.8a) of the nanoshaved region in the PVP film image a prominent, bright square appears. In the friction loop (Figure 6.8b), the variation in photodiode response indicates the PVP and the glass areas in the sample.

Post-Lab Questions

1. According to the data shown in Table 6.1, what is the recommended setpoint for performing AFM lithography?
   a. 30 nN
   b. 5 nN

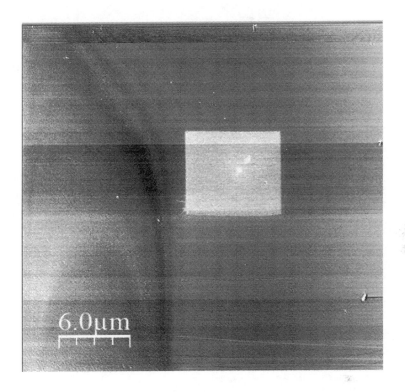

(a)

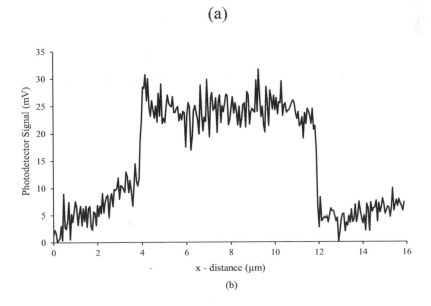

x - distance (μm)

(b)

FIGURE 6.8
Friction image of square etched in a PVP film using nanoshaving (a). Friction loop data for a
forward line scan across the pattern (b).

    c. 3 nN

    d. 10 nN

2. According to the data shown in Table 6.1, what is the recommended setpoint for imaging nanoshaved structures?

    a. 30 nN

    b. 5 nN

    c. 3 nN

    d. 10 nN

3. Based on the LFM image shown in Figure 6.8a, the bright region is likely due to:

    a. PVP film

    b. bare glass

    c. the tip losing contact with the sample surface

4. Based on the LFM image shown in Figure 6.8a, the bright region is likely due to:

    a. PVP film

    b. bare glass

    c. the tip losing contact with the sample surface

5. Based on the observations in the LFM image, the tip likely experienced higher torsional forces over the:

    a. PVP film

    b. bare glass

6. Based on the observations in the LFM image, the tip likely experienced lower torsional forces over the:

    a. PVP film

    b. bare glass

7. Based on the LFM data obtained in this experiment, which region is likely to have the highest coefficient of friction (μ)?

    a. PVP film

    b. bare glass

8. Based on the LFM data obtained in this experiment, which region is likely to have the lowest coefficient of friction (μ)?

    a. PVP film

    b. bare glass

9. In the friction loop data shown in Figure 6.8b, the region producing photodetector signals ≤ 15 mV is likely:

a. PVP film

b. bare glass

10. In the friction loop data shown in Figure 6.8b, the region producing photodetector signals Between 15–30 mV is likely:

a. PVP film

b. bare glass

End-of-Chapter Questions

1. True or False. AFM users cannot monitor topography and friction channels simultaneously.

   a. True

   b. False

2. The three main uses of LFM include:
   Choose all that apply.

   a. fundamental study of tribology

   b. characterization of heterogeneous materials

   c. determination of frictional properties of uniform materials

3. Which of the following nanotribological phenomena can users characterize with LFM?
   Choose all that apply.

   a. friction

   b. wear

   c. lubrication

4. In order to discern hydrophobic or hydrophilic regions in a molecular film with LFM:

   a. image with a higher setpoint

   b. image with a lower setpoint

   c. functionalize the AFM tip

   d. none of the above

5. True or False. Compositional mapping is possible with LFM.

   a. True

   b. False

6. What component of the AFM system monitors vertical bending and torsion of the cantilever?

    a. laser

    b. photodiode

    c. feedback loop

    d. tip

7. Which of the following cantilever motions provide information regarding the frictional properties of the sample surface?

    a. vertical deflection

    b. torsion

8. LFM is a mode that is available when the AFM system is operating in:

    a. contact mode

    b. tapping mode

9. Which of the following shows the correct relationship between frictional forces (F_f), normal forces (F_n) and the coefficient of friction (μ)?

    a. $F_f = \dfrac{\mu}{F_n}$

    b. $F_f = \dfrac{F_n}{\mu}$

    c. $F_f = \mu \cdot F_n$

    d. $F_f = \mu + F_n$

10. Which of the following equations is used to determine the friction signal produced by the four-quadrant photodiode?

    a. $(A + B) - (C + D)$

    b. $(A - B) + (C - D)$

    c. $(A + B) - (B + D)$

    d. $(A - B) + (B - D)$

11. The signal difference between the upper and the lower halves of the photodiode is the _____ deflection signal.

    a. lateral

    b. normal

12. The signal difference between the left and right halves of the photodiode is the _____ deflection signal.

    a. lateral

    b. normal

13. True or False. The lateral deflection signal produced by the four-quadrant photodiode will always be the same in the forward and reverse scan.

   a. True

   b. False

14. In friction loop data, forward and reverse traces are _____.

   a. the same

   b. different

15. All of the following are used to determine the frictional force (Ff) except:

   a. the normal spring constant

   b. the torsional spring constant

   c. the photodiode sensitivity

   d. the friction loop signal

References

[1] E. Meyer, "Atomic force microscopy," *Prog. Surf. Sci.*, vol. 41, pp. 3–49, 1992.

[2] R. Buzio and U. Valbusa, "Nanolubrication studied by contact-mode atomic force microscopy," in *Modern Research and Educational Topics in Microscopy*, A. Mendez- Vilas and J. Diaz, Eds., Badajoz, Formatex, 2007, pp. 491–499.

[3] J. Tamayo and R. Garcia, "Friction force microscopy characterization of semiconductor heterostructures," *Mater. Sci. Eng. B.*, vol. 42, pp. 122–126, 1996.

[4] P. Eaton and P. West, *Atomic Force Microscopy*, Oxford: Oxford University Press, 2010.

[5] R. Bennewitz, "Friction force microscopy," in *Fundamentals of Friction and Wear on the Nanoscale*, E. Gnecco and E. Meyer, Eds., Heidelberg, Springer, 2015, pp. 3–16.

[6] J. L. Wilbur, H. A. Biebuyck, J. C. MacDonald and G. M. Whitesides, "Scanning force microscopies can image patterned self-assembled monolayers," *Langmuir*, vol. 11, pp. 825–831, 1995.

[7] U. D. Schwarz and H. Holscher, "Atomic-scale friction studies using scanning force microscopy," in *Modern Tribology Handbook*, vol. 1, Boca Raton, CRC Press, 2001, pp. 641–660.

[8] K. D. Jandt, "Atomic force microscopy of biomaterials surfaces and interfaces," *Surf. Sci.*, vol. 491, pp. 303–332, 2001.

[9] G. Zavala, "Atomic force microscopy, a tool for characterization, synthesis, and chemical processes," *Colloid Polym. Sci.*, vol. 286, pp. 85–95, 2008.

[10] A. Shegaonkar, C. Lee and S. Salapaka, "Feedback scheme for improved lateral force measurement in atomic force microscopy," in *American Control Conference*, Seattle, 2008.

[11] "Nanotechnology/AFM," [Online]. Available: https://en.wikibooks.org/wiki/Nanotechnology/AFM#Lateral_Force_Microscopy. [Accessed 21 March 2019].

[12] S. Breakspear, J. R. Smith, T. G. Nevell and J. Tsibouklis, "Friction coefficient mapping using the atomic force microscope," *Surf. Interface Anal.*, vol. 36, pp. 1330–1334, 2004.

[13] H. Y. Nie, "Scanning probe techniques," [Online]. Available: http://publish.uwo.ca/~hnie/spm.html. [Accessed 21 March 2019].

[14] S. Xu and G. Y. Liu, "Nanometer-scale fabrication by simultaneous nanoshaving and molecular self-assembly," *Langmuir*, vol. 13, pp. 127–129, 1997.

[15] I. Horcas, R. Fernandez, J. M. Gomez-Rodriguez, J. M. Colchero, J. Gomez-Herrero and A. M. Baro, "WSXM: A software for scanning probe microscopy and tool for nanotechnology," *Rev. Sci. Instrum.*, vol. 78, pp. 013705-1–013705-8, 2007.

7

Conductive Atomic Force Microscopy

7.0 Key Objectives

- Gain familiarity with CAFM probe characteristics.
- Learn the basic operation of CAFM electronics.
- Become familiar with the various types of CAFM analyses.

7.1 CAFM Overview

M. P. Murrel of Cambridge University in 1993 first modified an AFM to perform current measurements [1]. Conductive atomic force microscopes (CAFMs) arose from the need to measure local currents flowing between an ultrasharp tip and a sample [1]. CAFM is a powerful current-sensing technique for the characterization conductivity on the nanoscale [2]. CAFM, also referred to in the literature as conductive probe AFM (CP-AFM), conductive scanning probe microscope (C-SPM), or conductive scanning force microscope (C-SFM), is an AFM that records the currents flowing at the tip/sample nanojunction [1]. CAFM allows the simultaneous characterization of topography and conductivity of samples [1]. With CAFM it is possible to obtain both surface topography and local electrical characterization with nanoscale resolution [2]. The CAFM set-p (Figure 7.1) is very similar to that of the standard AFM. There are three main differences: (1) the probe tip must be conductive, (2) requirement of a voltage source to apply a potential difference between the tip and the sample holder, and (3) requirement of a preamplifier to convert the current signal into voltages that can be read by the computer. In CAFM experiments, the sample is usually fixed on the sample holder using a conductive tape or paste, the most widely used being silver paints [1].

Current measurement occurs using a two-electrode configuration. The AFM tip, covered with a conductive layer, acts like a nano-electrode while

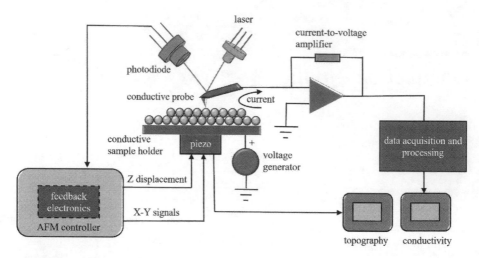

FIGURE 7.1
Conductive AFM (CAFM) setup.

the second electrode is a conductive plate beneath the sample [3]. CAFMs allow the user to select the value and polarity of the bias applied to the tip while keeping the sample holder grounded; therefore, the currents usually flow vertically through the sample [10]. The surface of the sample connects to the sample holder using silver paint or a wire bonder [4]. CAFM electronics involves the application of voltages ranging from −10 to +10 V. The observed currents fall between 1 pA and 10 µA [1]. The tip-sample current allows for conductance probing and mapping of samples with conducting and insulating areas or domains. This makes C-AFM useful for the study of electrically inhomogeneous substrates [5]. Conversion of the measured current to a voltage leads to the processing of the signal to quantify the electrical quantities [3].

7.2 CAFM Electronics

For nanoscale electric current measurements the proper design of the current-to-voltage amplifier is crucial due to the very small signals detected. Figure 7.2 shows the typical current-to-voltage amplifier typically used in CAFM. The amplifier enables conversion of measured currents to voltages using a feedback resistor, R_f [3]. Measured current signals flowing through the tip/sample nanojunction transform into digital voltages when they reach the preamplifier. The digital voltages reach the data acquisition (DAQ) card of the computer for reading [4].

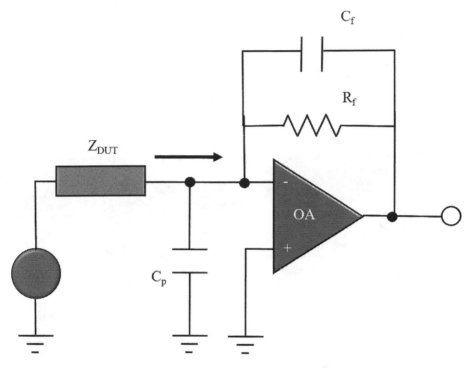

FIGURE 7.2
Current-to-voltage amplifier used in CAFM instruments.

7.3 CAFM Probe Characteristics

Reliable CAFM electrical measurements require a conductive tip. B-doped diamond- and PtIr-coated Si cantilevers prove to be suitable for CAFM imaging [6]. Conductive probes used in CAFM are standard AFM probes typically made of Si or Si_3N_4 coated with a thin conductive film [3, 7]. Generally, AFM tips coated with metal films fail after some time of use because the metal coating wears off [5]. The lateral resolution of the CAFM measurements depends on the effective area (A_{eff}), the conductivity of the sample, the scanning parameters (e.g. contact force), the radius of the tip, and the environmental conditions in which the experiments are performed [1]. The effective area, A_{eff}, is the sum of the spatial locations on the sample surface making contact with the CAFM tip [1]. The relationship between the effective area A_{eff} and the physical contact area A_c can vary based on the nature of the tip-sample interaction, as shown in Figure 7.3.

The most accepted method for the estimation of the physical contact area (A_c) between the CAFM tip and the sample is the Hertz contact theory. Figure 7.4 shows the tip-sample properties used in this theory. A_c

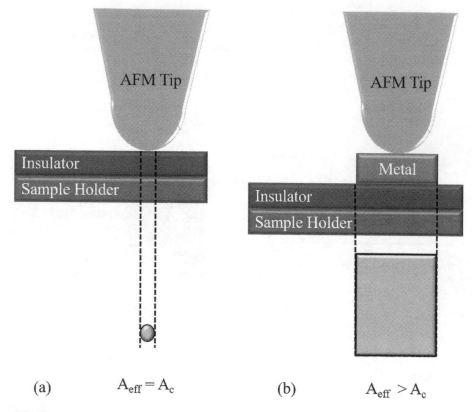

FIGURE 7.3
Relationship between A_{eff} and A_c when the conductive AFM tip makes contact with an insulator and a conductor.

determination (Equation 7.1) involves contact radius (r_c), the contact force (F_c), tip radius (R_{tip}), the modulus of elasticity (E_1, E_2), and the Poisson ratio (v_1, v_2) of the tip and sample. Review of the literature provides the parameters needed for Equation 7.1 [7].

$$A_c = \pi r_c^2 = \pi \left(F_c R_{tip} \frac{3}{4} \left(\frac{1-v_1^2}{E_1} + \frac{1-v_2^2}{E_2} \right) \right)^{\frac{2}{3}} \qquad (7.1)$$

7.4 Nanoscale Impedance Microscopy

Nanoscale impedance microscope (NIM) involves monitoring the magnitude and phase of the current at the driving frequency with either a

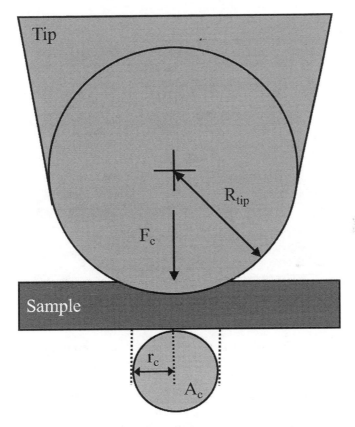

FIGURE 7.4
The tip-sample parameters used in the Hertz contact theory.

impedance analyzer or a lock-in amplifier (Figure 7.5) when an AC bias reaches the sample via a conductive tip. In this manner, NIM provides 10 nm spatial resolution maps of impedance variations [8]. This technique has revealed underlying electrolytic surface reactions, doping levels of semiconductors, and the properties of interfaces in organic and inorganic multilayer devices [8].

Impedance, Z, is the ratio between an applied voltage variation, V, and current response, I, as shown in Equation 7.2 [9].

$$Z(\omega) = V(\omega) / I(\omega) \qquad (7.2)$$

During NIM scans, measurement of the frequency dependent current transport involves application of an AC voltage between the tip and the substrate, resulting in the acquisition of the AC current flowing through the tip in contact with the surface. From the measured electrical current,

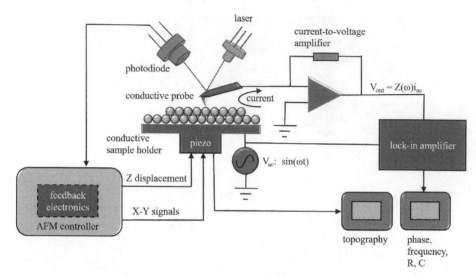

FIGURE 7.5
Schematic of a common nanoscale impedance microscopy.

the electrical impedance can be extracted [3]. The signal measured by the system depends critically on the nature of the tip/sample contact and thus the interpretation of AFM impedance data depends critically on understanding of the tip/sample interface [10]. Understanding impedance data involves the use of circuit models, the simplest of which is the ideal case of an RC circuit where a capacitor is in parallel with a resistor [9]. This model allows the extraction of material properties from impedance measurements [9]. Figure 7.6 illustrates a simple equivalent circuit model of the tip/sample contact. The contact is a parallel RC element with additional series resistors included to account for the tip and sample spreading resistances. The resistance of the tip (R_{tip}) is often significant given the small dimensions of AFM probes [10]. Here the equivalent circuit shows that the amplitude and phase differences across the interface yield voltage-dependent junction resistance R and capacitance C. The model also accounts for tip (R_{tip}) and sample (R_s) resistances [9].

7.4.1 Nanoscale Impedance Microscopy Data

NIM involves scanning the sample while a lock-in amplifier detects the modulated amplitude and phase of the cantilever response, which contains the local transport information. Using parameters associated with the equivalent circuit model (Figure 7.6) and Equation 7.3, phase information at the tip-sample interface can be determined [9].

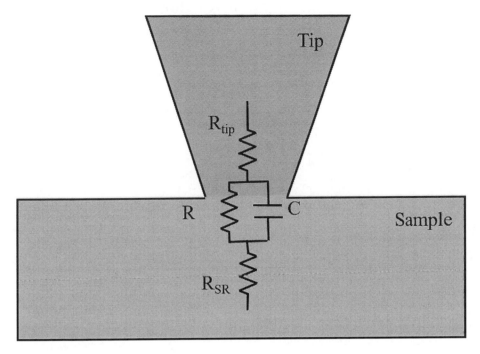

FIGURE 7.6
Equivalent circuit model used in nanoscale impedance microscopy.

$$\tan(\varphi_d) = \frac{\omega C_d R_d^2}{(R + R_d)R\omega^2 C_d^2 R_d^2} \qquad (7.3)$$

In Equation 7.3, φ_d is the phase at the interface boundary. R includes serial resistances such as spreading resistance and tip resistances. R_d and C_d are resistances and capacitances at the tip-sample interface, and ω is the frequency of the AC input signal. Here the lower frequency impedance response is due to resistive contributions while higher frequency responses are due to capacitive contributions [9]. By measuring frequency dependent phase and amplitude information, determination of resistance R_d and capacitance C_d values is possible [9]. Impedance data is illustrated graphically using Nyquist plots, with the imaginary impedance on the y-axis and the real impedance on the x-axis (Figure 7.7). Nyquist plots are generally in the form of arcs defined by a distinct time constant (τ) whose shape provides information on the resistive and capacitive nature of samples [9].

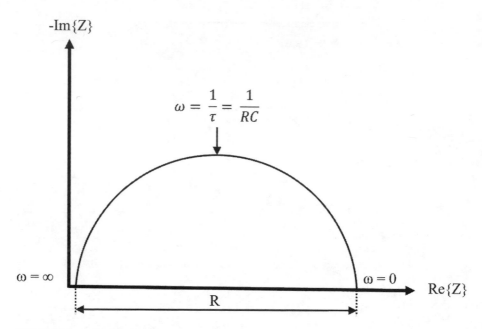

FIGURE 7.7
Impedance data presented in the form of a Nyquist plot.

7.5 Laboratory Exercise: CAFM Analysis of Silver Nanowires

7.5.1 Laboratory Objectives

The laboratory objectives covered in this laboratory exercises include:

- preparation of a silver nanowire sample for conductive imaging
- selecting the proper parameters for conductive imaging
- execution of AFM imaging to collect conductive and topographical images
- determination of electrical resistance using CAFM data and image processing software

7.5.2 Materials and Procedures

The suggested samples for use in this experiment are silver nanowires produced from the redox reaction between roughened or sputtered copper metal and dilute concentrations of silver nitrate. This silver nanowire synthesis is reported in the literature [11]. Use double-sided tape to mount the sample to a 2 cm imaging disk. In addition, silver paint and small-diameter electrical solder is needed to connect the sample to the CAFM system. The

AFM system used in this lab should have CAFM capabilities. Use a diamond scribe should to cut a 2 cm × 2 cm glass square from a clean microscope slide. Next, perform silver nanowires synthesis on the surface of the glass substrate according to instructions in the literature [11]. To mount the nanowire sample on a 2 cm imaging disc, use double-sided tape. Use silver paint and a strip of electrical solder to connect the sample to the CAFM base. Set up the CAFM for simultaneous conductive and topographical images according to the AFM manual. Use a bias voltage of 0.5 V for CAFM imaging.

7.5.3 Sample Data

Figure 7.8 shows a contact mode AFM image of a silver nanowire. This image was generated using free image processing software [12].

The blue marker in the topography image is the location used to produce the cross-sectional profile image (Figure 7.9). The cross-sectional profile image provides the physical dimensions of the nanowires.

Figure 7.10 shows a CAFM image of silver nanowires generated by free image processing software [13]. CAFM electrical data (current and resistance) at different lengths are shown in Table 7.1. Analysis of the CAFM data shows the increase in resistance with length (Figure 7.11) as reported in the literature [14]. Electrical resistance was determined using Ohm's law (Equation 7.4).

$$V = I \times R \qquad (7.4)$$

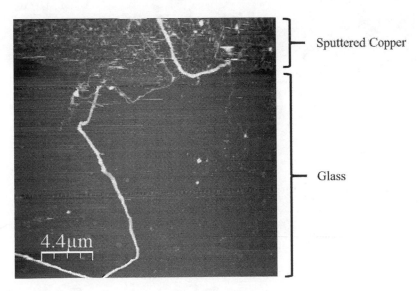

FIGURE 7.8
Contact mode image of silver nanowire.

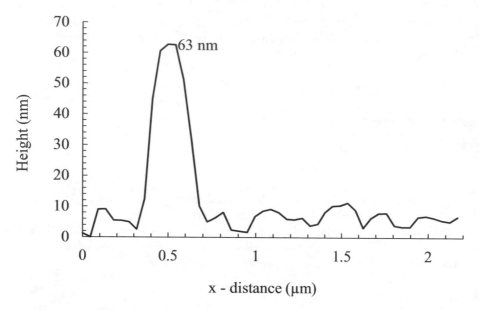

FIGURE 7.9
Cross-sectional profile of silver nanowire.

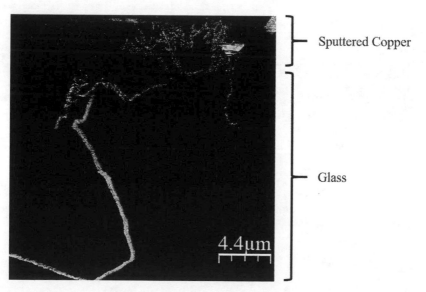

FIGURE 7.10
CAFM image of silver nanowire.

TABLE 7.1

Length and Resistance Measurements of Silver Nanowires Imaged with CAFM

Nanowire Length (µm)	Current (A)	Resistance (MΩ)
1.78	1.071×10^{-9}	466.85
4.62	8.46×10^{-10}	591.02
7.21	6.34×10^{-10}	788.64
17.36	5.2×10^{-10}	961.54
19.40	4.23×10^{-10}	1,182.03

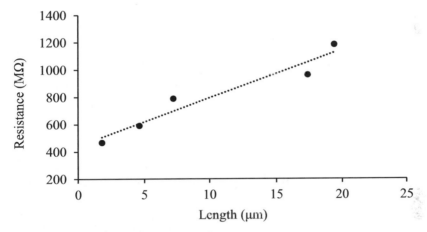

FIGURE 7.11
Linear relationship between electrical resistance and length.

Post-Lab Questions

1. True or False. AFM users cannot image topography and conductivity images simultaneously.
   a. True
   b. False

2. Which of the following image processing functions can be used to determine the physical dimensions of nanowires?
   a. leveling
   b. histogram adjust
   c. line profile
   d. filter

TABLE 7.2

Sample Nanowire Conductivity Data

Sample	Length (μm)
[1]	7.21
[2]	1.78
[3]	19.40
[4]	4.62

3. True or False. The physical dimensions of nanowires can be determined using conductivity images.

   a. True

   b. False

4. According to the conductivity image shown in Figure 7.10, conductive regions appear _____ and non-conductive regions appear _____.

   a. bright, dark

   b. dark, bright

5. Consider the following conductivity data shown in Table 7.2. Which nanowire is likely to possess the highest electrical resistance?

   a. [1]

   b. [2]

   c. [3]

   d. [4]

6. Based on the nanowire conductivity data listed in Table 7.2, which nanowire is likely to possess the smallest electrical resistance?

   a. [1]

   b. [2]

   c. [3]

   d. [4]

End-of-Chapter Questions

1. All of the following are requirements of a CAFM system except a:

   a. closed-loop scanner

   b. conductive probe

   c. voltage source

   d. current preamplifier

2. In order to perform conductivity imaging, the sample needs to be fixed on the sample holder using:

   a. double-sided tape

   b. silver paint

   c. transparent tape

   d. epoxy

3. True or False. The voltage bias used during CAFM imaging is fixed. Users cannot adjust the bias voltage.

   a. True

   b. False

4. Currents measured during CAFM imaging can be low as _____ and can reach _____.

   a. 1 mA, 10 A

   b. 1 µA, 10 mA

   c. 10 µA, 1 mA

   d. 1 pA, 10 µA

5. What is the role of the current preamplifier in CAFM systems?

   a. to amplify the bias voltage

   b. to supply voltage to the sample

   c. to convert measured currents to voltages

   d. to measure the voltage applied to the sample

6. Coated AFM tips fail after prolonged usage because:

   a. contaminants insulate the tip

   b. the bias voltage oxidizes the metal coating

   c. the coating wears off

   d. none of the above

7. All of the following have a dramatic impact on the resolution of conductive images except:

   a. scan size

   b. effective area (A_{eff})

   c. sample conductivity

   d. tip radius

8. True or False. DC voltages are used during nanoscale impendance microscopy (NIM) scans.
   a. True
   b. False

9. Which of the following are measured during NIM imaging?
   Choose all that apply.
   a. current magnitude
   b. phase of the current
   c. bias voltage
   d. electrical resistance

10. Which of the following components is used to monitor signals producing during NIM imaging?
   a. current amplifier
   b. lock in amplifier
   c. feedback loop
   d. laser

11. Understanding the impedance data from NIM imaging requires the use of:
   a. image processing
   b. equivalent circuit models

12. Tip-sample contact in NIM systems behaves similar to:
   Choose all that apply.
   a. resistors
   b. inductors
   c. capacitors
   d   relays

13. During NIM imaging, the resistance and capacitive nature of the tip-sample interface can be determined by measuring:
   Choose all that apply.
   a. phase
   b. amplitude
   c. bias voltage
   d. none of the above

14. The lower frequency impedance response measured using NIM is due to _____ contributions of the tip-sample interface.
   a. resistive
   b. capacitive

15. The higher frequency impedance response measured using NIM is due to _____ contributions of the tip-sample interface.

   a. resistive

   b. capacitive

References

[1] C. Pan, Y. Shi, F. Hui, E. Grustan-Gutierrez and M. Lanza, "History and status of the CAFM," in *Conductive Atomic Force MIcroscopy—Applications in Nanomaterials*, M. Lanza, Ed., Weinheim, Wiley-VCH Verlag GmbH & Co. KGaA., 2017.

[2] D. Mikulik, M. Ricci, G. Tutuncuoglu, F. Matteini, J. Vukajlovic, N. Vulic, E. Alarcon- Llado and A. F. Morral, "Conductive-probe atomic force microscopy as a characterization tool for nanowire-based solar cells," *Nano Energy*, vol. 41, pp. 566–572, 2017.

[3] L. Fumagalli, I. Casuso, G. Ferrari and G. Gomila, "Probing electrical transport properties at the nanoscale by current-sensing atomic force microscopy," in *Applied Scanning Probe Methods VIII: Scanning Probe Microscopy Techniques*, B. Bhushan, H. Fuchs and M. Tomitori, Eds., Heidelberg, Springer-Verlag, 2008, pp. 421–451.

[4] C. Pan, Y. Shi, F. Hui, E. G. Gutiererrez and M. Lanza, "History and Status of the CAFM," in *Conductive Atomic Force Microscopy: Applications in Nanomaterials*, M. Lanza, Ed., Weinheim, Wiley-VCH, 2017, pp. 1–28.

[5] J. Y. Park, S. Maier, B. Hendriksen and M. Salmeron, "Sensing current and forces with SPM," *Mater. Today*, vol. 13, pp. 38–45, October 2010.

[6] J. Alvarez, I. Ngo, M. E. Gueunier-Farret, J. P. Kleider, L. Yu, P. R. Cabarrocas, S. Perraud, E. Rouvière, C. Celle, C. Mouchet and J. P. Simonato, "Conductive-probe atomic force microscopy characterization of silicon nanowire," *Nanoscale Res. Lett.*, vol. 6, pp. 1–9, 2011.

[7] O. Krause, "Fabrication and reliability of conductive AFM probes," in *Conductive Atomic Force Microscopy: Applications of Nanomaterials*, M. Lanza, Ed., Weinheim, Wiley- VCH Verlag GmbH and Co., 2017, pp. 29–43.

[8] L. S. C. Pingree and M. C. Hersam, "Bridge-enhanced nanoscale impedance spectroscopy," *Appl. Phys. Lett.*, vol. 87, pp. 233117–233119, 2005.

[9] S. S. Nonnenmann, X. Chen and D. A. Bonnell, "High sensitivity scanning impedance microscopy and spectroscopy," in *Scanning Probe Microscopy for Energy Research*, D. A. Bonnell and S. V. Kalinin, Eds., Hackensack, NJ, World Scientific Publishing Pte. Ltd., 2013, pp. 457–476.

[10] R. O'Hayre, G. Feng, W. D. Nix and F. B. Prinz, "Quantitative impedance measurement using atomic force microscopy," *J. Appl. Phys.*, vol. 96, pp. 3540–3549, 2004.

[11] W. C. Sanders, P. D. Ainsworth, D. M. Archer, Jr., M. L. Armajo, C. E. Emmerson, J. V. Calara, M. L. Dixon, S. T. Lindsey, H. J. Moore and J. D. Swenson, "Characterization of micro- and nanoscale silver wires synthesized using a single-replacement reaction between sputtered copper metal and dilute silver nitrate solutions," *J. Chem. Educ.*, vol. 91, pp. 705–710, 2014.

[12] I. Horcas, R. Fernandez, J. M. Gomez-Rodriguez, J. M. Colchero, J. Gomez-Herrero and A. M. Baro, "WSXM: A software for scanning probe microscopy and tool for nanotechnology," *Rev. Sci. Instrum.*, vol. 78, pp. 013705-1–013705-8, 2007.

[13] D. Necas and P. Klapetek, "Gwyddion: An open-source software for SPM data analysis," *Cent. Eur. J. Phys.*, vol. 10, pp. 181–188, 2012.

[14] J. S. Roh, "Conductive yarn embroidered circuits for systems on textiles," in *Wearable Technologies*, J. H. Ortiz, Ed., London, InTech, 2018, pp. 161–174.

8

Oscillating Modes of AFM

8.0 Key Objectives

- Gain familiarity with forces associated with oscillating modes of AFM.
- Become familiar with the signals detected in oscillating modes of AFM.
- Understand probe behavior during oscillating modes of AFM.
- Understand the role of the feedback loop and the lock-in amplifier.
- Become familiar with the principles of phase imaging.

8.1 Tapping Mode

Tapping mode involves alternating tip-surface interactions. Synonymous terms include intermittent contact mode [1], dynamic mode, and AC mode [2]. During tapping mode, an external piezo actuator drives the oscillation of a microscale cantilever (Figure 8.1). The frequency of oscillation is the resonance frequency of the cantilever. During oscillation, the tip makes periodic contact with the sample during each cycle of oscillation [3–5]. Several approach and retract cycles occur at each pixel location [2]. The tip mounted at the end of the cantilever gently "taps" the sample surface during the bottom swing of the cantilever [6].

As the tip oscillates alternatively contacting the surface and lifting off, the amplitude oscillation generally ranges from 1–100 nm [5]. Large oscillation amplitudes overcome adhesion forces between the tip and sample surface [2]. Measurement of frequency shifts and changes in vibration amplitude occur during tapping mode imaging (Figure 8.2) [5].

As the tip interacts with the sample surface, attenuation of the vibration amplitude occurs. This attenuation makes it possible to monitor changes in sample height [7]. When the tip starts feeling the interaction with the surface, there a decrease in the vibration amplitude with decreasing tip-sample distance

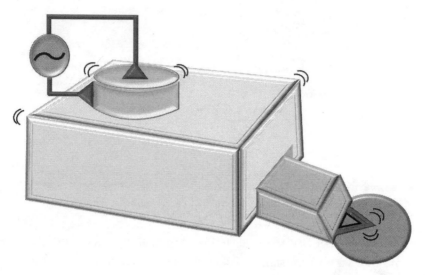

FIGURE 8.1
Cantilever oscillation of a cantilever using an external piezo actuator.

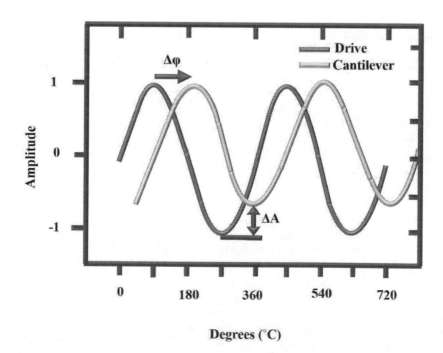

FIGURE 8.2
Differences in vibration amplitude and phase between the drive and measured signals.

(Figure 8.3) [4]. Variations in the vibration amplitude and resonant frequency provide information regarding sample topography and material properties [3].

The tip-surface contact time is small, and as a result, lateral forces are reduced considerably [6]. Tapping mode overcomes issues related to friction, adhesion, and electrostatic forces. This technique avoids dragging the tip across the surface, minimizing sample damage [7]. The elimination of continuous tip-sample contact that would otherwise produce substantial frictional forces damaging to the tip and the sample, makes tapping mode well-suited for imaging of weakly immobilized and soft samples such as polymers and thin films [6, 8]. In addition, the minimization of the destructive lateral forces allows the study of weakly adsorbed molecules on a substrate [9] and the study of compliant materials [10].

8.1.1 Tapping Mode Operation

Tapping mode AFM systems include a cantilever oscillation subsystem, deflection detection, signal conversion, a sample positioning subsystem,

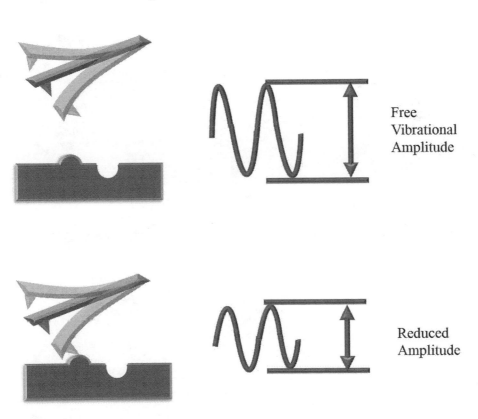

Free Vibrational Amplitude

Reduced Amplitude

FIGURE 8.3
Reduction of vibration amplitude due to tip-sample interactions.

and a feedback controller subsystem. As stated in a previous section, cantilever oscillation occurs using a cantilever/tip assembly driven mechanically by a piezoelectric transducer [3]. Cantilever deflection detection utilizes a position-sensitive photodiode, current-to-voltage electronics, preamplifiers, and low-pass filters to measure the dynamic, vertical bending of the cantilever. The sample positioning subsystem contains a piezoelectric tube scanner and a precision stepping motor allowing the positioning of the AFM probe in the x, y, and z positions along the sample surface [3]. An oscillator produces the cantilever drive signal (ω_{drive}). This signal drives the piezoelectric transducer at the cantilever base. Cantilevers have resonance frequencies that reach up to several hundred kHz [4]. To oscillate cantilevers at their resonance frequencies, it is necessary for the piezoelectric transducer to have a higher resonance frequency. Oftentimes this is not possible because the piezo tubes in the scanner have very low resonance frequencies [4]. For this reason, an additional piezo plate with a high resonance frequency, known as the dither piezo element, oscillates the cantilever base. The beam deflection method allows for the monitoring of dynamic cantilever deflection (Figure 8.4) [4].

When the signal from the position-sensitive photodiode reaches the preamplifier electronics, conversion of the signal to a voltage proportional to the dynamic cantilever deflection occurs. This signal is proportional to the

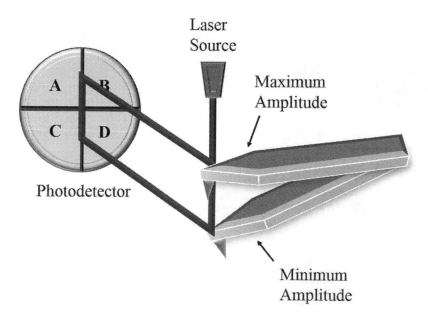

FIGURE 8.4
Detection of cantilever oscillation using a position-sensitive photodetector.

cantilever oscillation amplitude [4]. As the tip scans across the sample, the feedback control signal provides an accurate representation of the sample topography [11].

8.1.2 Mechanical Properties of Tapping Mode Tips

The "resonance frequency" of the oscillator refers to the driving frequency at which vibration amplitude (A) is maximized [12]. The motion of a tapping mode cantilever follows the classic equation for the damped and driven simple harmonic oscillator [12]. The solution to this equation describes steady-state oscillation driven by an applied force and includes a phase shift δ (Equation 8.1) [12]:

$$z(t) = A \cos(\omega t - \delta) \tag{8.1}$$

where phase shift is described using Equation 8.2:

$$\delta = \tan^{-1}\left(\frac{\omega\beta}{\omega_0^2 - \omega^2}\right) \tag{8.2}$$

given that

ω_0 = natural frequency of the cantilever

ω = measured frequency of the cantilever

$\beta = b/m$, where b is the damping coefficient (coefficient of friction) and m is the reduced mass

Additionally, Equation 8.3 describes the vibration amplitude of the oscillating cantilever [12]:

$$A = \frac{D}{\sqrt{(\omega_0^2 - \omega^2)^2 + \omega^2\beta^2}} \tag{8.3}$$

8.1.3 Tapping Mode Forces

During tapping mode imaging, tip-sample forces involve long-range attractive forces and short-range repulsive forces (Figure 8.5) [3]. Tip-sample forces include van der Waals force, dipole-dipole interactions, and electrostatic forces. These forces act on the cantilever resulting in a decrease in the amplitude of cantilever oscillation [13].

It is also important to mention that tip-sample forces consist of capillary forces (due to a water layer coating both the cantilever and the sample) and repulsive forces (due to the contact between the tip and the sample) (Figure 8.6).

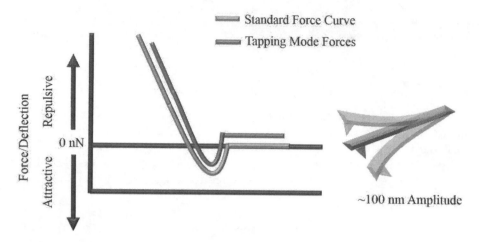

FIGURE 8.5
Force-distance relationship between the tip and sample during tapping mode imaging.

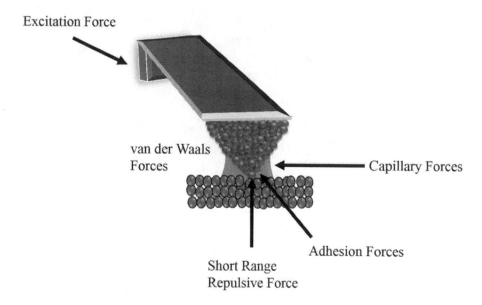

FIGURE 8.6
Tip-sample forces present during tapping mode AFM images.

As the tip approaches the sample, tip-sample forces are exclusively van der Waals forces. Next, capillary forces influence the tip due to interactions between thin films of water that cover the sample and the tip under ambient humidity. A connective column of liquid is established. When the tip finally makes contact with the sample, repulsive forces are present. When the tip retracts from the sample, the capillary force continues until the connective column of liquid is broken [14]. Interactions between the AFM tip and sample surface result in a decrease in vibration amplitude and phase [15]. Attractive tip-sample forces yield lower resonance frequencies, and repulsive tip-sample forces yield higher resonance frequencies [4, 5].

8.1.4 Tapping Mode Parameters

There are several parameters associated with tapping mode AFM. These parameters include the resonance frequency of the free cantilever, the measured frequency of the cantilever, cantilever spring constant, the quality factor of the cantilever (Q), the free vibration amplitude of the cantilever, the frequency (phase) shift, and the setpoint [4]. The resonance frequency of the free cantilever and the cantilever spring constant are inherent properties of the AFM probe [4]. Careful consideration of cantilever stiffness is imperative [16]. Cantilevers that are too stiff result in sample and or tip destruction. Cantilevers that are too soft may not interact with the sample enough to generate sufficient contrast or may fail to stay in contact with the sample surface. The free vibration amplitude is the amplitude of cantilever oscillation when the cantilever is vibrating in free space above the sample [16]. This parameter is set in units of voltage. Rough samples generally require larger free vibration amplitudes. Setpoint expresses a percentage of the free vibration amplitude maintained during imaging. Lower setpoints result in a more aggressive tip-sample interaction [16].

8.1.5 Q Factor

Air friction and spring losses dampen tip and cantilever motion. For this reason, a damping term in necessary. The quality factor (Q) quantifies the amount of damping. For weakly damped systems, such as AFM tips vibrating in air, a practical determination of Q involves dividing the resonance frequency of the cantilever by the width of the resonance peak (where the width is taken at the points where the amplitude is equal to $1/2$ of the amplitude, $\Delta\omega$) (Figure 8.7) [4].

A probe with a high Q factor will take longer for the AFM system to respond to abrupt changes in the sample topography (Figure 8.8a); therefore, slower scan speeds are required for an accurate trace of sample topography. To improve the resolution of the image while using higher scan speeds, a cantilever with a lower Q factor is required (Figure 8.8b) [11]. However, reducing the cantilever Q factor to increase scan speed will ultimately increase the tip-sample force. This leads to an increased risk of damage to the sample [11]. Most AFM cantilevers designed for tapping mode have a higher Q factor than is needed for the desired resolution. However, to increase the scan speed, it is desirable to decrease the *Q* factor of the cantilever [11].

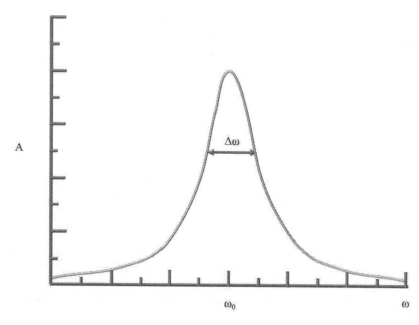

FIGURE 8.7
Graphical determination of the Q factor.

8.1.6 Feedback Loop Operation in Oscillating Modes

During tapping mode AFM, the feedback system detects shifts in resonance frequencies and changes in vibration amplitude (Figure 8.9) [5]. It is possible to select amplitude or phase shift to keep constant via the feedback loop [15]. The most commonly measured signal in tapping mode is the vibration amplitude, a value kept constant via the feedback loop [4], [17], [18]. The feedback loop matches the measured amplitude to the setpoint value by changing the z position of the sample [4]. Raising or lowering the z piezo adjusts the vibration amplitude to restore the original setpoint value [7]. As the feedback loop maintains a constant vibration amplitude throughout a scan, the feedback signal corresponds to a height profile [4]. Thus processing the z feedback signal leads to the production of topographical images [4].

8.1.7 Lock-In Amplifier

Vibration amplitude measurement involves detection with a lock-in amplifier; lock-in amplifiers detect and measure very small AC signals [4]. Lock-in amplifiers measure the oscillation amplitude at the reference frequency, and filters signals at other frequencies [1]. The output of the lock-in amplifier is the input signal for the feedback controller [4]. In a typical lock-in amplifier setup, the device under test (DUT) reads a sinusoidal signal (Figure 8.10a).

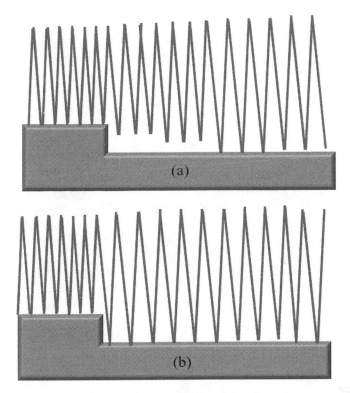

FIGURE 8.8
Responses to surface topography with a cantilever possessing a higher Q factor (a) and a lower Q factor (b).

The device response $V_s(t)$ is known as the input signal and is used in addition to the reference signal $V_r(t)$ by the lock-in amplifier to determine the amplitude (R) and phase (θ). Determination of the amplitude and phase requires the use of a dual-phase demodulation circuit (Figure 8.10b) [19]. This circuit splits the input signal into two signals; one is the reference signal and a 90° phase-shifted copy of it. The signals pass through low-pass filters to reject the noise and convert signals into amplitude R and the phase θ components [19]:

8.2 Non-Contact Mode

Non-contact mode involves imaging when the tip is in close proximity (within a few nanometers of), but not in contact with, the sample surface [20]. The tip oscillates above the sample during scanning [6]. The oscillating tip

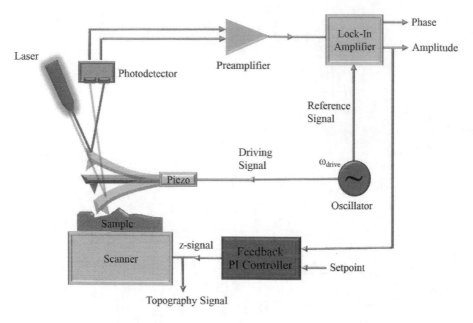

FIGURE 8.9
AFM feedback system for tapping mode.

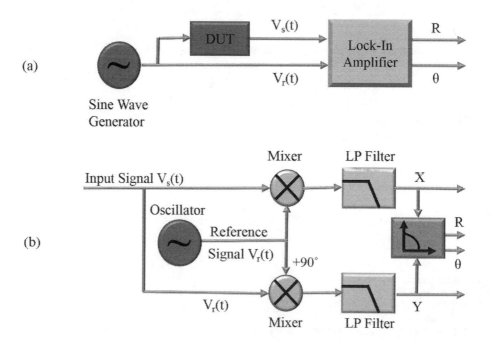

FIGURE 8.10
Standard lock-in amplifier (a). Lock-in amplifier with a demodulator circuit (b).

hovers at approximately 1 nm above the sample surface [20]. The primary advantage of non-contact mode is that the tip is always in the weak attractive region (Figure 8.11), so the sample and tip are not easily damaged [1]. The forces exerted on the samples are low (10^{-12} N) [6]. Due to this very delicate interaction, imaging of difficult samples such as gels and nanoparticles weakly adhered to substrate is possible. Additionally, imaging films without tearing, plowing, puncturing, or other deleterious effects is possible [2]. Also, this mode is capable of imaging of biological and polymeric samples without alteration of surfaces [15]. However, there are some disadvantages associated with non-contact mode. These disadvantages include decreased resolution due to the presence of a contaminant layer on the sample surface. Additionally, non-contact mode usually requires ultra-high vacuum (UHV) to achieve optimal imaging [6]. Just like tapping mode, non-contact mode requires a lock-in amplifier, an amplifier that measures the signal at the reference frequency and filters signals at other frequencies [1].

8.2.1 Non-Contact Mode Forces

The predominate forces existing between the tip and sample surface are van der Waals forces. Additionally, other forces are possible such as electrical and magnetic forces. However, if the surface is not charged and if the surface and tip are not ferromagnetic, the predominant forces are van der Waals [1]. The oscillating tip, held a few nanometers above the sample surface, only senses attractive forces because it does not reach the surface, which would otherwise generate repulsive forces [2]. It is important to point out that the resolution of non-contact mode imaging decreases with increasing tip-sample distance [20]. The attractive forces felt between the tip and sample modifies the resonance behavior thus, the tip's oscillation amplitude decreases and enables the tracking of topography [2].

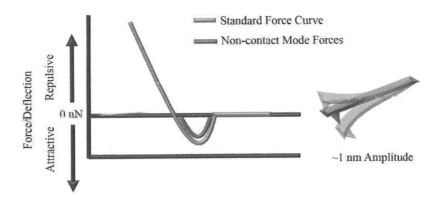

FIGURE 8.11
Force-distance relationships between the tip and sample during non-contact mode imaging.

8.3 Phase Imaging

There also exists a property-sensitive imaging mode, known as phase imaging. In phase imaging, the phase of the cantilever vibration relative to the signal sent to the cantilever's piezoelectric driver is simultaneously monitored and recorded by the AFM controller [10]. Phase imaging is suitable for the collection of information about surface properties, not detectable by other AFM modes. Phase imaging is useful for the identification of physical domains with different composition and material properties. In addition, phase imaging allows recognizing regions with different elasticity and or other structural heterogeneities, such as crystalline/amorphous composition of polymers [21]. Phase imaging applications also include identification of contaminants, mapping different components in composite materials, and mapping electrical and magnetic properties. The latter has wide-ranging implications in data storage and semiconductor industries [5]. Phase imaging plays a promising role in the study of material properties at nanoscale [10]. The change in cantilever resonant frequency is in response to additional force gradients. Attractive forces make the cantilever "softer,"

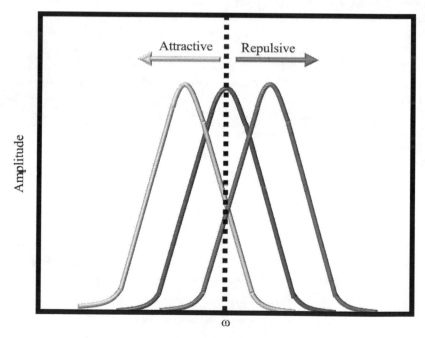

FIGURE 8.12
Effect of tip-sample forces on cantilever resonant frequency.

effectively reducing the measured frequency. In contrast, repulsive forces make the cantilever "stiffer," effectively increasing the measured frequency (Figure 8.12) [10].

The phase difference (phase lag) detected between the driven oscillations of the cantilever and the measured oscillations results from differences in material properties (Figure 8.13) [6]. For example, a flat polymer sample containing two different polymers with varying stiffness at the surface may be distinguishable in the AFM phase mode, although the two polymers would not be distinguishable from the AFM surface topography image [20]. Phase lag monitoring occurs simultaneously with

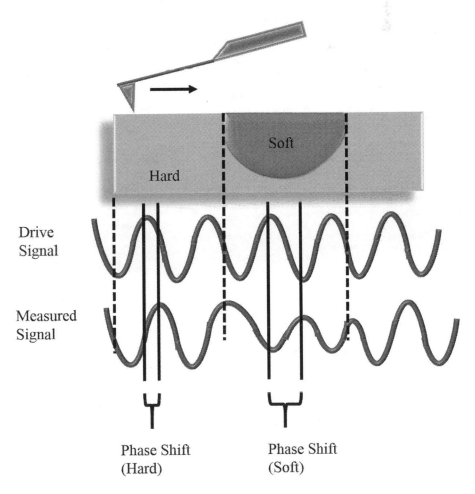

FIGURE 8.13
The effect of material properties on phase angle.

amplitude by the control electronics. The electronics record and transform the data into AFM images [9]. It is possible to conduct phase imaging in tandem with topographic imaging while operating AFM systems
in tapping mode [9].

An excited cantilever oscillation will exhibit a phase shift (φ) between the
drive and the response, as defined by Equation 8.13:

$$d = A\sin\left(2\pi ft + \varphi\right) \tag{8.13}$$

where d = deflection; A = amplitude; f = frequency; t = time; and φ = phase
shift [16].

The phase shift is 90° when the probe is far from the sample and there is
no interaction between the AFM tip and the sample. The AFM phase image
appears dark when the phase shift decreases and appears bright when the
phase shift increases [10]. Over the regions with a higher elastic modulus,
repulsive forces predominate. More damping occurs over the more viscous
regions. Hence, the areas with higher elastic moduli segments appear bright
and the more viscous samples appear dark [10].

8.4 Laboratory Exercise: Phase Imaging of Metal/Polymer Nanostructures

8.4.1 Laboratory Objectives

The objectives of this experiment include:

- fabrication and metallization of polymeric nanostructures
- acquisition of phase images
- qualitative assessment of phase images

8.4.2 Materials and Procedures

A 2 cm × 2 cm glass square cut from a clean microscope slide using a diamond
scribe is a sufficient substrate. For best results, use the silver nanostructure
synthesis reported in the literature to prepare samples for the experiment
[22]. Use double-sided tape to mount the sample to a 2 cm imaging disk. The
AFM system used in this lab should have the ability to acquire phase images
while operating in tapping mode. Set up the AFM for tapping mode and
phase imaging according to the AFM manual. Table 8.1 list recommended
imaging parameters for the laboratory exercise.

TABLE 8.1

Recommended Tapping Mode Parameters for Metal/Polymer Imaging

Parameter	Value
Resonant Frequency	170 kHz
Free Vibration Amplitude	245 mV
Setpoint	35%
Scan Size	20 µm

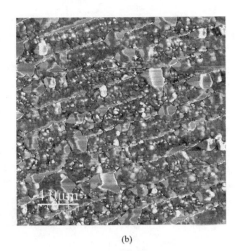

(a)   (b)

FIGURE 8.14
Topography image (a) and phase image (b) of electrodeposited silver nanostructures on PVA.

8.4.3 Sample Data and Results

Figures 8.14a and 8.14b show topography and phase images of polymer/metal nanostructures respectively. The phase image (Figure 8.14b) show bright and dark areas, indicating different components in the nanostructures. The results shown in the phase AFM data are consistent with the moduli reported in the literature; silver possess a modulus of 82.5×10^3 megapascals (82.5 gigapascals) [23] and the reported polyvinyl alcohol (PVA) modulus is 6.4 megapascals [24]. As stated in Section 8.3.1, areas with a higher elastic modulus appear brighter than areas that are more viscous [10]; the phase image of the electrodeposited nanostructure is consistent with this observation. The brighter regions show areas consisting of electrodeposited silver and the dark regions show areas consisting of polymer.

Post-Lab Questions

1. Which of the following parameters are used in tapping mode imaging only?
   Choose all that apply.

   a. setpoint

   b. resonant frequency

   c. scan speed

   d. vibration amplitude

2. True or False. Users cannot obtain topography and phase images simultaneously during tapping mode imaging.

   a. True

   b. False

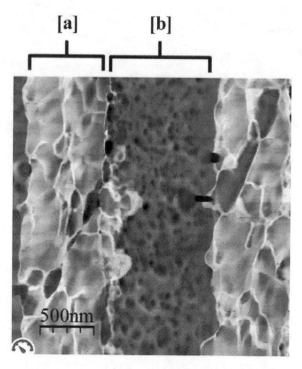

FIGURE 8.15
A sample phase AFM image.

3. Bright regions in phase AFM images represent areas with _____ modulus values. Dark regions in phase AFM images represent areas with _____ modulus values.

   a. low, high

   b. high, low

4. Consider the sample phase AFM image shown in Figure 8.15. Which area is likely to have a higher modulus of elasticity (more rigid)?

   a. [a]

   b. [b]

5. Consider the sample phase AFM image shown in Figure 8.15. Which area is more viscous (softer)?

   a. [a]

   b. [b]

End-of-Chapter Questions

1. All of the following are terms used to describe tapping mode imaging except:

   a. modulation mode

   b. intermittent contact mode

   c. dynamic mode

   d. AC mode

2. The amplitude oscillation range for tapping mode is:

   a. 1–5 nm

   b. 1–10 nm

   c. 1–50 nm

   d. 1–100 nm

3. Attenuation of which of the following allows is used to determine sample height during tapping mode imaging?

   a. phase

   b. current

   c. amplitude

   d. frequency

4. The small tip-sample contact that occurs during tapping mode results in the reduction of which of the following?
Choose all that apply.

   a. friction

   b. adhesion

   c. electrostatic forces

   d. sample damage

5. True or False. A separate piezo element is needed to oscillate the cantilever.

   a. True

   b. False

6. The resonance frequency refers to:

   a. the frequency of tip scanning

   b. the frequency with the highest vibrational amplitude

   c. the frequency with the highest tip-sample forces

   d. none of the above

7. Which of the following forces are exerted between the tip and sample during tapping mode imaging?
Choose all that apply.

   a. van der Waals

   b. dipole-dipole forces

   c. electrostatic forces

   d. magnetic forces

8. True or False. Capillary forces decrease the resolution of images produced during tapping mode imaging.

   a. True

   b. False

9. The free vibration amplitude is the amplitude of the cantilever oscillation when the cantilever:

   a. makes contact with the sample surface

   b. taps a hard sample

   c. taps a soft sample

   d. vibrates above the sample surface

10. Which type of sample generally requires larger free vibrational amplitudes?

a. rough

b. smooth

11. When tapping mode is used, the setpoint refers to:

    a. the forces exerted between the tip and sample

    b. the magnitude of capillary forces

    c. the percentage of the free vibration amplitude maintained

    d. the maximum vibrational amplitude

12. A more aggressive tip-sample interaction occurs when _____ setpoints are used during imaging.

    a. lower

    b. higher

13. The probe characteristic that describes the degree of damping is the:

    a. resonance frequency

    b. Q factor

    c. vibrational amplitude

    d. spring constant

14. _____ scans are required for an accurate trace of sample topography for tips with a high Q factor.

    a. Slower

    b. Faster

15. _____ scans are required for an accurate trace of sample topography for tips with a low Q factor.

    a. Slower

    b. Faster

16. The most commonly used feedback signal in tapping mode AFM is:

    a. phase shift

    b. vibration amplitude

    c. cantilever deflection

    d. resonance frequency

17. During tapping mode imaging, the feedback loop raises or lowers the z piezo to maintain constant:

    a. phase shift

    b. vibration amplitude

   c. cantilever deflection

   d. resonance frequency

18. Which of the following components is used to monitor signals during tapping mode imaging?

   a. photodiode

   b. laser

   c. lock-in amplifier

   d. current-to-voltage amplifer

19. The smallest tip-sample force exerted between the tip and sample during tapping mode imaging is:

   a. 10^{-3} N

   b. 10^{-6} N

   c. 10^{-9} N

   d. 10^{-12} N

20. During non-contact mode imaging, the AFM tip:

   a. makes direct contact with the sample surface

   b. gently taps the sample surface

   c. hovers above the sample surface

   d. none of the above

21. The separation between the AFM tip and the sample surface is:

   a. 100 nm

   b. 50 nm

   c. 10 nm

   d. 1 nm

22. The predominate forces that exerted between the AFM tip and sample surface during non-contact mode imaging are:

   a. magnetic forces

   b. van der Waals forces

   c. electrostatic forces

   d. repulsive forces

23. During phase imaging, the measured frequency is _____ when soft samples are scanned.

   a. reduced

   b. increased

24. During phase imaging, the measured frequency is _____ when stiffer samples are scanned.

    a. reduced
    b. increased

25. During phase imaging, more damping of the vibration amplitude occurs when the oscillating tip is above _____ samples.

    a. viscous
    b. rigid

26. In phase images, areas with higher elastic moduli appear _____.

    a. bright
    d. dark

27. In phase images, areas with lower elastic moduli appear _____.

    a. bright
    d. dark

28. The phase shift is _____ when the oscillating probe does not interact with the sample surface.

    a. 180°
    b. 90°
    c. 45°
    d. 0°

References

[1] G. Zavala, "Atomic force microscopy, a tool for characterization, synthesis, and chemical processes," *Colloid Polym. Sci.*, vol. 286, pp. 85–95, 2008.

[2] G. Haugstad, *Atomic Force Microscopy—Understanding Basic Modes and Advanced Applications*, Hoboken: John Wiley and Sons, 2012.

[3] Y. J. Wang, "Constant force feedback controller design using PID-like fuzzy technique for tapping mode atomic force microscopes," *Intell. Control Automat.*, vol. 4, pp. 263–279, 2013.

[4] B. Voigtlander, *Scanning Probe Microscopy—Atomic Force Microscopy and Scanning Tunneling Microscopy*, Heidelberg: Springer, 2015.

[5] H. J. Butt, B. Cappella and M. Kappl, "Force measurements with the atomic force microscope: Technique, interpretation, and applications," *Surf. Sci. Rep.*, vol. 59, pp. 1–152, 2005.

[6] R. De Oliveira, D. Albuquerque, T. Cruz, F. Yamaji and F. Leite, "Measurement of the nanoscale roughness by atomic force microscopy: Basic principles and applications," in *Atomic Force Microscopy—Imaging, Measuring, and Manipulating Surfaces at the Atomic Scale*, V. Bellitto, Ed., Croatia, InTech, 2012, pp. 147–174.

[7] N. Hernandez-Pedro, E. Rangel-Lopez, B. Pineda and J. Sotelo, "Atomic force microscopy in detection of viruses," in *Atomic Force Microscopy Investigations Into Biology—From Cell to Protein*, Rijeka, InTech, 2012, pp. 235–252.

[8] A. F. Sarioglu and O. Solgaard, "Time-resolved tapping-mode atomic force microscopy," in *Scanning Probe Microscopy in Nanoscience and Nanotechnology 2*, B. Bhushan, Ed., 2011, Heidelberg, Springer, pp. 3–37.

[9] P. J. James, M. Antognozzi, J. Tamayo, T. J. McMaster, J. M. Newton and M. J. Miles, "Interpretation of contrast in tapping mode AFM and shear force microscopy—A study of Nafion," *Langmuir*, vol. 17, pp. 349–360, 2001.

[10] F. Z. Fang, Z. W. Xu and S. Dong, "Study on phase images of a carbon nanotube probe in atomic force microscopy," *Meas. Sci. Technol.*, vol. 19, pp. 1–7, 2008.

[11] M. W. Fairbairn, S. Moheimani and A. J. Fleming, "Q control of an atomic force microscope microcantilever: A sensorless approach," *J. Microelectromech. Syst.*, vol. 20, pp. 1372–1381, 2011.

[12] R. M. Overney and M. Sarikaya, "Intermittent-contact mode force microscopy and Electrostatic Force Microscopy (EFM)," in *NUE Unique SPM Workshop: Nanoscience on the Tip*, Seattle, 2010.

[13] S. Chatterjee, S. S. Gadad and T. K. Kundu, "Atomic force microscopy—A tool to unveil the mystery of biological systems," *Resonance*, pp. 622–642, July 2010.

[14] M. H. Korayem, A. Kavousi and N. Ebrahimi, "Dynamic analysis of tapping-mode AFM considering capillary force interactions," *Sci. Iran.*, vol. 18, pp. 121–129, 2011.

[15] U. Maver, T. Maver, Z. Persin, M. Mozetic, A. Vesel, M. Gaberscek and K. Stana-Kleinschek, "Polymer characterization with the atomic force microscope," in *Polymer Science*, F. Yilmaz, Ed., Croatia, InTech, 2013, pp. 113–132.

[16] "Tapping modes," 2018. [Online]. Available: www.nanosurf.com/en/support/afm-modes#dynamic. [Accessed 7 August 2018].

[17] F. Moreno-Herrero, J. Colchero, J. Gomez-Herrero and A. M. Baro, "Atomic force microscopy contact, tapping, and jumping modes for imaging biological samples in liquids," *Phys. Rev. E*, vol. 69, pp. 031915-1–031915-9, 2004.

[18] Z. Instruments, "Principles of lock-in detection and the state of the art," 2016.

[19] B. Bhushan and O. Marti, "Scanning probe microscopy—Principle of operation, instrumentation, and probes," in *Nanotribology and Nanomechanics I: Measurement Techniques and Nanomechanics*, B. Bhushan, Ed., Berlin, Springer, 2011, pp. 37–110.

[20] K. D. Jandt, "Atomic force microscopy of biomaterials surfaces and interfaces," *Surf. Sci.*, vol. 491, pp. 303–332, 2001.

[21] M. Marrese, V. Guarino and L. Ambrosio, "Atomic force microscopy: A powerful tool to address scaffold design in tissue engineering," *J. Funct. Biomater.*, pp. 1–20, 2017.

[22] W. C. Sanders, G. Johnson, R. Valcarce, P. Iles, H. Fourt, K. Drystan, D. Edwards, J. Vernon, S. Ashworth, A. Barucija and Z. Curtis, "Electrodeposition of silver micro- and nanoscale wires in the capillaries of PDMS stamps modified with hydrophilic polymers," *J. Chem. Educ.*, vol. 96, pp. 1218–1223, 2019.

[23] D. R. Smith and F. R. Fickett, "Low-temperature properties of silver," *J. Res. Natl. Inst. Stand. Technol.*, vol. 100, pp. 119–171, March–April 1995.

[24] S. M. M. Dadfar, G. Kavoosi and S. M. A. Dadfar, "Investigation of mechanical properties, antimicrobial features, and water vapor permeability of polyvinyl alcohol thin films reinforced by glutaraldehyde and multiwalled carbon nanotube," *Polym. Compos.*, vol. 35, pp. 1–8, 2014.

9

Image Processing

9.0 Key Objectives

- To become familiar with the following image processing techniques:
  - leveling
  - histogram adjust
  - fast Fourier transform
  - line profile

9.1 Introduction

AFM rarely provides information that can undergo analysis without any further data processing. Generally, basic leveling procedures need to be implemented for the removal of sample tilt. Additionally, AFM data is susceptible to tip-sample convolution. These issues are addressed by utilizing image processing software [1]. Usually, algorithms used for data processing are provided by the AFM manufacturer and are therefore proprietary, preventing AFM data from opening with other graphics software [1]. Applications available for open-source image processing software include, false-color representation, 3D data display, color maps, selection of multiple channels, non-planarity compensation, distances in profile graph, FFT filtering, and correction of image defects [1].

9.2 Leveling

During AFM scans, variations from between line scans (i.e., changes in the average height, tilts, or low-frequency noise) are difficult to avoided (Figure 9.1a). For this reason, image processing functions such as leveling (flatten) are very common when analyzing SPM images. Leveling works by subtracting a function from each scan line in the raw AFM data file (Figure 9.1b). The simplest function is a zero-order function: the average of each line. Other common possibilities are a first-order function (a straight line) or a second-order function (a parabola). Leveling filters can eliminate slopes, bows, and/or bands in the images coming from low-frequency noise [2]. Leveling does possess limitations. Leveling procedures can produce image artifacts. For this reason, a feature exclusion procedure is often required [3]. If a line is flat, and the equation is subtracted from the data, the line data will remain uniform. When a line contains very different heights and the same process is used, the originally lower parts will be even lower. This causes "artificial" negative contrast regions (shadowing effects) around the higher parts of the filtered image. Image processing software packages have standard tools to circumvent these issues: discard regions, draw master paths, etc. [2]. Most image processing software packages allow users to exclude areas during the leveling procedure. When an area is excluded, it is not used for the calculation of the background in the image [4]. Leveling is usually the first image processing function carried out on AFM data [5]. In many instances, leveling is the only processing function performed on AFM data. Leveling is important because substrates with substantial tilt will mask the changes in height associated with the sample of interest. Even minimal substrate tilt can have deleterious effects [3]. Another issue circumvented using the leveling function is scanner bow. Scanner bow is a direct result of the

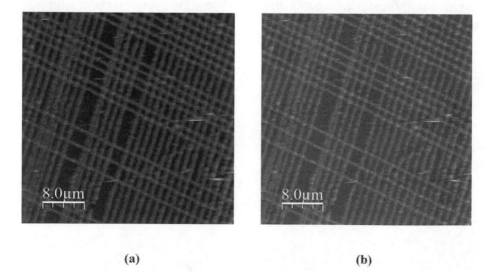

(a)   (b)

FIGURE 9.1
AFM image before leveling (a) and after leveling (b).

swinging motion of the free end of the scanner. This action introduces curves in the terminal region of AFM images [3].

9.2.1 Polynomial Fitting

A method association with leveling is polynomial fitting. This involves fitting each line of the AFM data to a polynomial equation. Then the polynomial is subtracted from the scan line; this action flattens the line scan and centers the height data to zero. In most cases, this action is applied to horizontal line scans because horizontal scans are usually the fast scan axes. The polynomial order of the equation can vary from 0th to 3rd order [3]. The leveling function uses 0th, 1st, 2nd, or 3rd order equations to remove tilt from each line scan [6]. The AFM user selects the appropriate best-fit polynomial; the polynomial is then subtracted from the AFM data [6]. Line-by-line polynomial leveling is the simplest method for removing unwanted background bow and tilt from AFM images [4]. Polynomial order and the effects on AFM images is summarized in Table 9.1. Table 9.2 lists the equations applied to AFM scan lines during leveling [6].

In many instances, a first order leveling procedure will suffice. However, if evidence of scanner bow is present in the AFM image, a second order leveling procedure is necessary to remove the scanner bow effects [5].

TABLE 9.1

Leveling Order Effects on AFM Images

Polynomial	Image Effect
0	Height offset of each scan line is set to the same value.
1	A straight line equation is subtracted from each scan line, and an offset is applied.
2	A quadratic equation is subtracted from each scan line, and an offset is applied.
3	A third order polynomial is subtracted from each scan line, and an offset is applied.

TABLE 9.2

Equations Applied to AFM during Leveling

Polynomial Order	Equation
0	$z = a$
1	$z = a + bx$
2	$z = a + bx + cx^2$
3	$z = a + bx + cx^2 + dx^3$

Note: a = Offset and b = Slope

9.3 Histogram Adjust

Histogram adjust involves analyzing a histogram containing plots of pixel height versus frequency (Figure 9.2). Each height is assigned a color value. AFM software packages will expand the available colors over the entire range of the z scale. As users perform histogram adjust on AFM data, users will observe that 90% or more of the data is confined into a narrow region of the histogram. Small amounts of outlying data points (i.e., very low or very high parts of the topography) may reflect real topographical features; however, this data can be attributed to errors or glitches in the data. When performing a histogram adjust, users may decide to reject upper and/or lower points in an effort to better distribute the color scale over the majority of the height data. The histogram adjust function can substantially increase the contrast in the majority of the image, and often helps greatly to visualize finer details in the image (Figure 9.3a-b). To enhance the contrast in an AFM image, users simply stretch the color scale across the relevant peak in the histogram. It is important to note that histogram adjust is very sensitive to the quality of the leveling. Poor leveling will bring peaks in the histograms closer together [7]. Histogram adjust function allows users to distribute a broad range of colors across a color scale. The histogram adjust

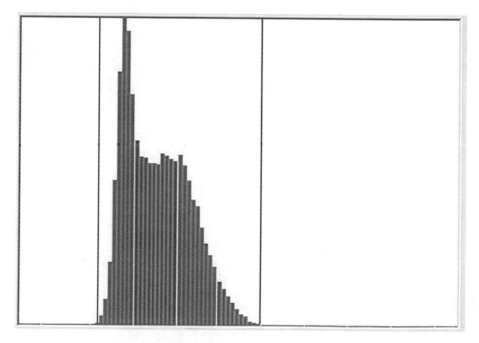

FIGURE 9.2
Histogram used to adjust the contrast of AFM images.

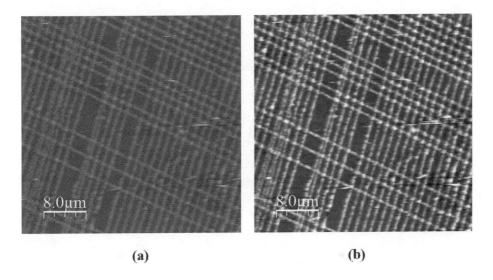

(a)   **(b)**

FIGURE 9.3
AFM images before (a) and after (b) histogram adjust.

function is used to show surface features that are not visible when the color scale takes the entire Z range of the image. After a histogram adjust is performed, features can be visualized and measured in an image that could not be observed in the original image [4].

9.4 Filtering

9.4.1 External Vibrations

Ambient vibrations found in the room where the AFM is located can result in vibrational artifacts in an image. The artifacts appear as oscillations in the image. Acoustic and floor vibrations can result in image artifacts [4]. Floor vibrations can oscillate up and down several μm, with frequencies below 5 Hz [4]. If not removed, these vibrations produce periodic structures in AFM images. Very flat samples are susceptible to these type of vibrations [4].

9.4.2 Fast Fourier Transform

Often there is unwanted high and low frequency noise that appears in AFM images. This noise can be removed by filtering [4]. A 2-D Fourier transform converts AFM data from the spatial domain to the frequency (or wavelength) domain [5]. This procedure involves a mathematical operation referred to as the fast Fourier transform; for this reason, this procedure is often called the FFT analysis. When FFT analyses are conducted on AFM data, images appear in terms of wavelength, or frequency. This assists in the identification of repeating patterns in images [5]. Ambient noises possess a characteristic frequency. For this reason, ambient noise is easily identified and removed using the FFT analysis [5]. Fourier filtering takes an image and calculates its frequency components, called the FFT image. Unwanted frequency components are identified and removed from the FFT image. When the FFT image is back transformed, the resulting AFM image will not have the frequency components that were removed in the FFT image. FFT filtering is particularly effective on images with repetitive patterns [4].

9.5 Line Profiles

The extraction and measurement of line profiles is one of the most frequently used post-processing analysis technique [5]. Line profiles are used to measure dimensions in AFM images. Post-processing software with line profile options allow users to arbitrarily define lines in AFM images to measure distances. These lines can be horizontal, vertical, or drawn at any angle. After the line is drawn, the post-processing software constructs a plot with the distance along the arbitrary line along the x-axis, and the sample height along the y-axis [5]. The plot allows AFM users to observe feature heights and widths [5].

End-of-Chapter Questions

1. Leveling is used to correct for:
   Choose all that apply.

   a. scanner bow

   b. low-frequency periodic noise

   c. sample tilt

   d. high-frequency periodic noise

2. Oftentimes, _____ is the first and only image processing step performed on AFM images.

   a. histogram adjust

   b. line profile

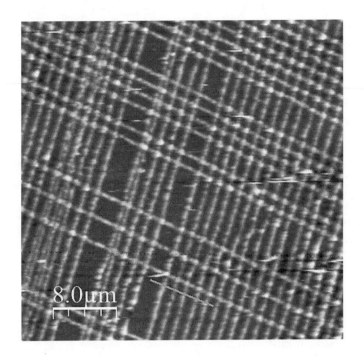

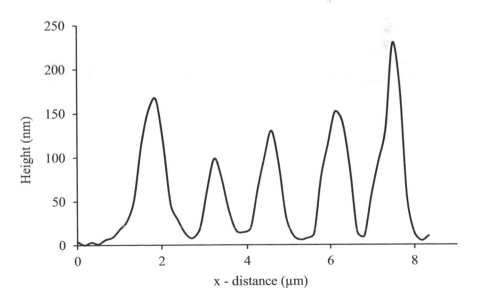

FIGURE 9.4
Topographical AFM image (a). Line profile generated from topographical AFM data.

    c. leveling

    d. filter

3. Which of the following polynomial equations are subtracted from AFM line scans during the leveling function?
Choose all that apply.

    a. 0th order

    b. 1st order

    c. 2nd order

    d. 3rd order

4. Which of the following image processing functions is used to improve contrast in AFM images?

    a. leveling

    b. line profile

    c. filter

    d. histogram adjust

5. Histogram adjust allows AFM users to:

    a. remove high-frequency periodic noise from images

    b. remove low-frequency periodic noise from images

    c. assign colors to heights in an AFM image

    d. remove the effects of scanner bow from AFM images

6. Which of the following image processing function allows finer details in samples to be revealed?

    a. line profile

    b. histogram adjust

    c. leveling

    d. filter

7. Which of the following is responsible for producing periodic noise in AFM images?

    a. high setpoint values

    b. low setpoint values

    c. external vibrations

    d. a and c

8. Which of the following is used to remove periodic noise from AFM images?

    a. leveling

    b. FFT

   c. line profile

   d. histogram adjust

9. During fast Fourier transform (FFT) image processing, _____ data is converted to _____ data.
Choose all that apply.

   a. spatial, wavelength

   b. wavelength, spatial

   c. spatial, frequency

   d. frequency, special

10. Which of the following image processing functions is used to determine distances in AFM images?

   a. histogram adjust

   b. leveling

   c. line profile

   d. filter

11. An AFM image contains evidence of scanner bow. Which of the following can remove the effects of scanner bow from the image?

   a. 1st order leveling

   b. 2nd order leveling

   c. FFT

   d. line profile

12. Nanoparticles with 30 nm diameters are difficult to see in an AFM image due to sample tilt. Which of the following image processing functions should be used to address this issue?

   a. Histogram adjust

   b. Leveling

   c. FFT

   d. Line profile

13. Oscillations from a nearby electromechanical pump produce periodic ripples in AFM images. Which of the following image processing functions should be used to address this issue?

   a. Histogram adjust

   b. Leveling

   c. FFT

   d. Line profile

References

[1] D. Necas and P. Klapetek, "Gwyddion: An open-source software for SPM data analysis," *Cent. Eur. J. Phys.*, vol. 10, pp. 181–188, 2012.

[2] A. Gimeno, P. Ares, I. Horcas, A. Gil, J. M. Gomez-Rodriguez, J. Colchero and J. Gomez-Herrero, "'Flatten plus': A recent implementation in WSxM for biological research," *Bioinformatics*, vol. 31, pp. 2918–2920, 2015.

[3] "Biomedical engineering reference-leveling," The-Crankshaft Publishing, [Online]. Available: http://what-when-how.com/Tutorial/topic-55iija/Atomic-Force-Microscopy-114.html. [Accessed June 2010].

[4] P. E. West, *Introduction to Atomic Force Microscopy: Theory, Practice, Applications*, Signal Hill: P. West, 2006.

[5] P. Eaton and P. West, *Atomic Force Microscopy*, Oxford: Oxford University Press, 2010.

[6] "Flatten," Bruker, 2010. [Online]. Available: www.nanophys.kth.se/nanophys/facilities/nfl/afm/icon/bruker-help/Content/SoftwareGuide/Offline/ModifyCommands/Flatten.htm. [Accessed June 2019].

[7] "Biomedical engineering reference—Histogram adjust," The-Crankshaft Publishing, [Online]. Available: http://what-when-how.com/Tutorial/topic-55iija/Atomic-Force-Microscopy-119.html. [Accessed June 2019].

Index